INNOVATIVE ENERGY CONVERSION FROM BIOMASS WASTE

INNOVATIVE ENERGY CONVERSION FROM BIOMASS WASTE

Arif Darmawan
Agency for the Assessment and Application of Technology (BPPT), Puspiptek Serpong, Tangerang Selatan, Indonesia

Muhammad Aziz
Institute of Industrial Science, The University of Tokyo, Meguro-ku, Tokyo, Japan

ELSEVIER

Elsevier
Radarweg 29, PO Box 211, 1000 AE Amsterdam, Netherlands
The Boulevard, Langford Lane, Kidlington, Oxford OX5 1GB, United Kingdom
50 Hampshire Street, 5th Floor, Cambridge, MA 02139, United States

Library of Congress Cataloging-in-Publication Data
A catalog record for this book is available from the Library of Congress

British Library Cataloguing-in-Publication Data
A catalogue record for this book is available from the British Library

ISBN: 978-0-323-85477-1

For information on all Elsevier publications visit our website at https://www.elsevier.com/books-and-journals

Publisher: Candice Janco
Acquisitions Editor: Peter Adamson
Editorial Project Manager: Michelle Fisher
Production Project Manager: Prem Kumar Kaliamoorthi
Cover Designer: Christian Bilbow

Typeset by TNQ Technologies

Contents

Contributors

Muhammad W. Ajiwibowo
Department of Mechanical Engineering, University of Indonesia, Depok, Indonesia

Muhammad Aziz
Institute of Industrial Science, The University of Tokyo, Meguro-ku, Tokyo, Japan

Muhammad Kunta Biddinika
Faculty of Industrial Technology, Universitas Ahmad Dahlan, Indonesia

Arif Darmawan
Agency for the Assessment and Application of Technology (BPPT), Puspiptek Serpong, Tangerang Selatan, Indonesia

Baskoro Lokahita
Department of Transdisciplinary Science and Engineering, Tokyo Institute of Technology, Midori-ku, Yokohama, Kanagawa, Japan

Koji Tokimatsu
Department of Transdisciplinary Science and Engineering, Tokyo Institute of Technology, Midori-ku, Yokohama, Kanagawa, Japan

CHAPTER 1

An overview of biomass waste utilization

Arif Darmawan[1] **and Muhammad Aziz**[2]
[1]Agency for the Assessment and Application of Technology (BPPT), Puspiptek Serpong, Tangerang Selatan, Indonesia; [2]Institute of Industrial Science, The University of Tokyo, Meguro-ku, Tokyo, Japan

1.1 Introduction: energy, sustainability, and efficiency

Fossil fuels largely dominate today's global energy supply. Fossil fuel is combustible deposits, including coal, natural gas, crude oil, and heavy oils, are made from decomposition of animals and plants and buried over hundreds of millions of years. The massive use of fossil fuels, especially by direct combustion, has caused environmental problems, such as air pollution and global temperature increase. When fossil fuels are burned, they release greenhouse gases (carbon dioxide and nitrous oxide) to the environment. The rise in greenhouse gas concentration can amplify the greenhouse effect by trapping extra heat in the atmosphere, causing the Earth's temperature to rise. Other pollutants are also released (sulfur dioxide and nitrogen oxides) during combustion, contributing to acid rain and various health problems primarily related to human respiratory function. Another side effect of fossil fuel use is water pollution due to contamination in energy recovery sites and raw fossil fuel extraction and processing. This contamination can have devastating impacts on human life. However, it seems a gradual transition is the best possible way to leave fossil fuels in the ground. Reducing fossil fuels also requires alternative energy sources that offer a sustainable supply with lower environmental impacts.

Ideally, sustainable energy development consists of three main conditions: promoting renewable energy, low environmental impact, and achieving high process efficiency. Renewable energy refers to the energy produced from sources that naturally can be replenished. The major types of renewable energy sources are hydropower, geothermal, wind power, biomass, and solar. The interest in renewable energy as an alternative source is increasing. Many countries have also been promoting new policies and considerable investments to increase renewable energy share in the national

Innovative Energy Conversion from Biomass Waste
ISBN 978-0-323-85477-1
https://doi.org/10.1016/B978-0-323-85477-1.00007-5

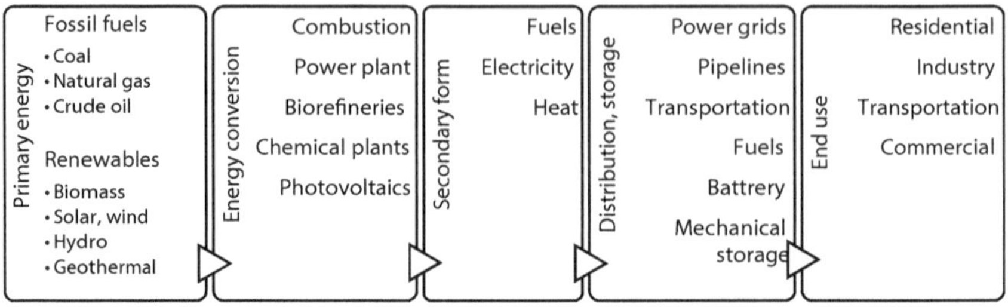

Figure 1.1 Schematic of energy flows from primary energy supply to final energy use.

energy mix. Both renewables and fossil fuels as primary energy sources are usually not suitable for direct use by end users. These primary sources must be converted into common secondary energy forms like electricity, heat, or fuels (diesel, gasoline, hydrogen, etc.) before being either stored or distributed to end users (see Fig. 1.1). The primary to secondary energy form's conversion process takes place in various plants such as fossil fuel and renewables power plants, biorefineries, chemical processing plants, or heat recovery by direct combustion. In the last step, the secondary energy form is distributed to provide energy services in residential, industry, transportation, and commercial sectors.

Among the available types of renewable sources, biomass is a sustainable alternative to replace fossil fuels. Unlike other primary sources, biomass is more evenly distributed throughout the world. Biomass is considered CO_2 neutral if it can be managed sustainably; therefore, it is crucial to improve the current bioenergy systems. Biomass resources can be classified into three types: natural growth, energy crops, and biomass waste. The utilization of energy crops and biomass waste can avoid potential conflict with food production or prevent a potential shortage of biomass-based commodities. As discussed earlier, biomass feedstock as primary energy is not suitable for direct use by final consumers. They should be utilized to produce secondary energy forms or energy carriers such as biofuels, heat, or power through various conversion technologies. Biofuel products can further be classified according to states of matter as solid (densified biocoal), liquid (such as ethanol, methanol, biodiesel, dimethyl ether), and gaseous (such as hydrogen, biomethane).

Innovative integrated systems to efficiently generate electricity or produce biofuels are required, considering biomass waste's high potential as an energy source. Process efficiency simply means the ratio between the useful output and the input into a process. In the past, efficiency is usually evaluated using mass and energy balance. However, this approach has the main disadvantage since the quality of materials and energy flows is not involved, and no environmental reference is used. Eventually, the concept of exergy was born based on first (energy) and second (entropy) thermodynamics law [1]. Exergy, a measure of energy quality, is the maximum amount of work obtainable when a material or process flow is brought to equilibrium with its environment. Thermodynamic indicators based on exergy are commonly accepted as the most natural way to measure the performance of different energy technology processes, chemical engineering, transportation, agriculture, and so on.

Efficiency improvement of biomass conversion can be performed in a design strategy using the pinch technology approach. The pinch technology's basic idea is to correctly perform process integration by considering all the heat transfers occurring in a system to reduce energy consumption. The method has been adopted in several studies employing biomass feedstocks for ethylene and ethanol production [2,3], hydrogen production [4], biodiesel production [5] including power generation [6]. However, these studies mostly focused on heat exchanger networks instead of manipulating the system to make it more efficient; in other words, it is only considering temperature as a quality parameter.

The book's main task is to propose and discuss a suitable method of efficiency improvement by adopting exergy recovery and process integration technologies. The basic idea of the approach is performing exergy elevation in a single process before being integrated with other processes via a heat exchanger network. This idea is substantially different from that of the pinch technology. An onion diagram can be drawn to show the overall process design steps (Fig. 1.2). A design process is performed from the inside to the outside of the "onion." The core of the whole process is the reaction layer. Since we focus on an integrated energy system for biomass utilization, the reactions involved are mainly conversion processes such as drying (pretreatment), combustion, gasification, steam cycle, combined cycle, and chemical looping. After a process development, such as physical separation and reactor design, we can determine the possibility of performing exergy recovery in each process, including developing the heat exchanger network (for process integration) and the requirements for heating and cooling.

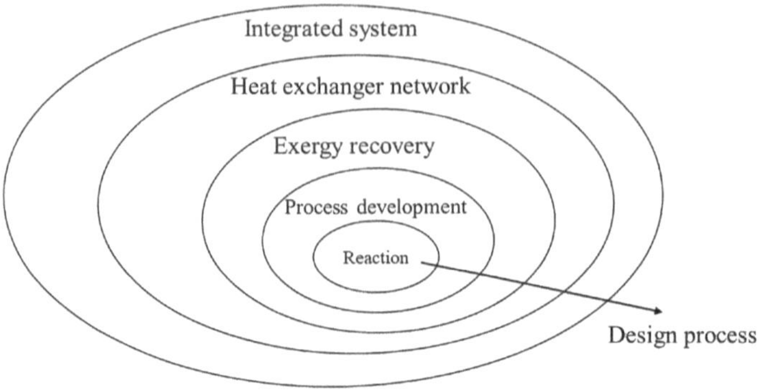

Figure 1.2 An onion diagram of design strategy adopting exergy recovery and process integration.

1.2 Global energy situation

Due to population and global economic growth, energy consumption is increasing rapidly. The world's population is expected to grow from 7.8 billion in 2020 to more than 9.5 billion in 2050 according to the UN scenario [7,8]. Humans consume energy for transportation, heating/cooling, or electricity in their daily activities. However, the rise of energy consumption is not uniform among countries because of several factors such as the economy and geographical situation, energy policy, emission goal, available energy supply, etc. Currently, the world's energy supply is mainly dependent on fossil fuels. In 2017, over 70% of global energy demand growth was met by fossil fuel resources. In 2018, oil contributed the highest share of about 34% or 4662 Mtoe of global energy supply, followed by coal (27%) and natural gas (24%) as shown in Fig. 1.3 [9]. In developing countries, fossil fuels are still subsidized, resulting in a heavy dependence on it and inefficient usage.

The United States is the top oil producer with an average of 17.87 million b/d, accounting for 18% of the world's oil production in 2018. Saudi Arabia is the second-largest producer with 12.42 million b/d (12%), followed by Russia with an average of 11.4 million b/d (11%), Canada with an average of 5.27 million b/d in 2018 (5%), and China with an average 4.82 million b/d (5%) [10]. China and the United States were also the highest net oil importer, as reported in 2019, with 22.6% and 12.5% of overall imported crude oil, respectively, followed by India (9.7%), Japan (6.9%), Korea (6.6%), and other countries. China, the top coal-producing country since 1985, reporting 3550 Mt of coal production in 2018 [11].

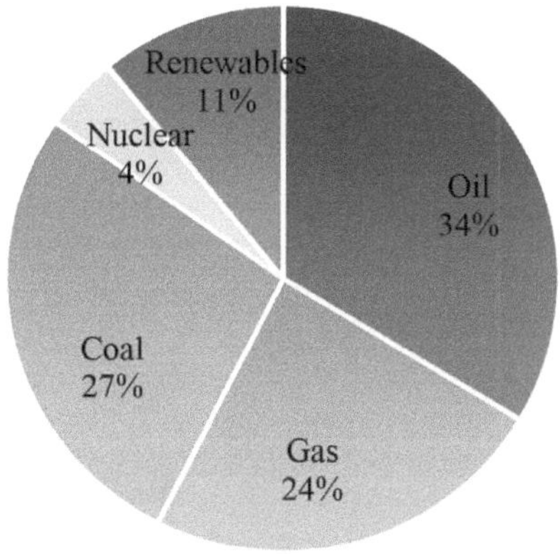

Figure 1.3 Fuel shares in total primary energy supply (TPES), 2018.

Although among the highest coal producers, China and India were listed as the top five coal importers together with Japan, South Korea, and Taiwan. Other top fossil fuel-producing countries are shown in Table 1.1 [12,13].

Fig. 1.4 shows the world's total final energy consumption by fuel 2001–17. In the last 20 years, the final energy consumption was steadily rising except in 2009 (fell by 1.1%) due to the global economic recession. It was recorded as the first decline since 1982. According to the IEA, the total world final energy consumption was 9717 Mtoe, with the share of fossil fuels being over 80% in 2017 [14]. With 37 member countries, the OECD consumes a relatively large proportion of about 38.19% of the total final energy consumption [14]. In 2018, two-third of the overall increase in the world's energy demand was dominated by China, the United States, and India combined. Those three countries were also top energy consumers, followed by Russia, Japan, South Korea, Germany, Canada, and Brazil.

Table 1.1 The World's top fossil fuel producers in 2018.

No	Oil	Million barrel/day	Coal	Million tonnes	Natural gas	Billion cubic meters
1	United States	17.87	China	3550	United States	864
2	Saudi Arabia	12.42	India	764	Russia	741
3	Russia	11.4	United States	684	Iran	232
4	Canada	5.27	Australia	502	Canada	188
5	China	4.82	Indonesia	474	Qatar	168

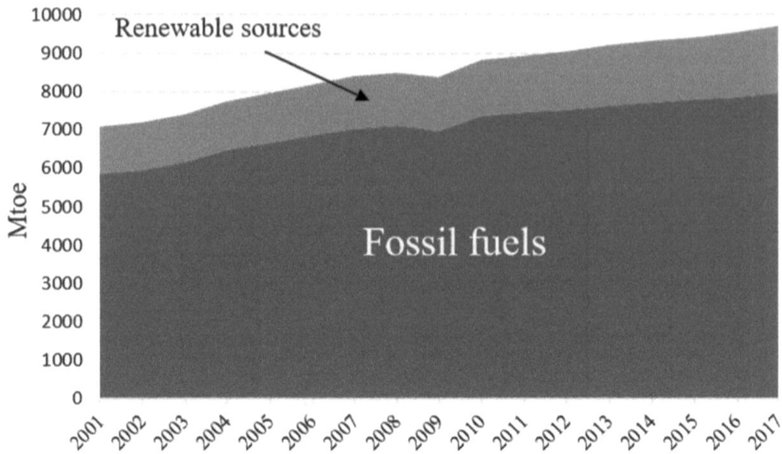

Figure 1.4 World total final energy consumption by fuel 2001–17.

Furthermore, about 18% of the total energy consumption was from renewable sources, including traditional biomass. The traditional biomass used for heating and cooking in developing countries accounted for 7.5% of the total energy consumed. The most significant portion of the renewable share (nontraditional biomass) was renewable thermal energy including biomass, solar, and geothermal heat (4.2%), followed by hydropower (3.6%), other renewable power sources including wind power and solar PV (2%), and transport biofuels (1%) [15]. As a result of a large portion of fossil fuels, energy-related CO_2 emission increased by 1.7% to 33.1 Gt in 2018, the fastest growth of carbon emissions in seven consecutive years. The power and heat generation sector accounted for over 40% of the total emitted CO_2, followed by transportation, industry, and others. Coal use in power alone contributed about 10 Gt CO_2 (30% of total emitted CO_2), mostly in Asia. Another issue is high inequality within and among countries; 13% of the world population or 940 million people still live without electricity in 2016 [10].

As of 2017, energy consumed for heating and cooling has the highest portion, followed by transportation and power with 51%, 32%, and 17%, respectively (see Fig. 1.5). Heating and cooling use thermal energy for some activities such as water heating, space heating/cooling, residential cooling/heating, refrigeration, and industrial processing like drying and cooking. It includes any energy other than electricity used for motive power in any applications [15]. It was estimated that the share of total renewable energy (excluding traditional biomass use) in the heating and cooling was about

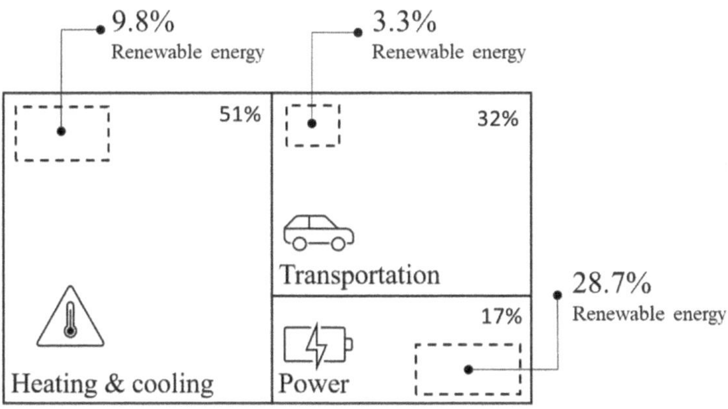

Figure 1.5 Renewable energy in total final energy consumption by sector, 2016.

9.8%. Although the use of fossil fuel (mostly oil) accounts for the biggest portion of energy consumption in the transportation sector, the share of biofuels (mainly ethanol and biodiesel) increased by about 18% between 2013 and 2017.

Many countries such as New Zealand, Germany, Costa Rica, Fiji, France, Spain, and United Kingdom have national targets to increase renewable energy use and become carbon-neutral by 2050. Sweden's parliament also voted to cut all greenhouse gas emissions by 2045 [16,17]. To achieve those commitments, they must find ways to satisfy national energy demands by gradually replacing fossil fuels. The European Union (EU) members set their target to reach 20% total share of renewables in the power sector by 2020 and at least 32% by 2030 [18]. Among the 28 EU members, 12 countries have already reached their target to increase renewables share, including Bulgaria, Czechia, Denmark, Estonia, Greece, Croatia, Italy, Latvia, Lithuania, Cyprus, Finland, and Sweden. The highest share of renewables in 2018 was recorded by Sweden with more than half (54.6%), followed by Finland (41.2%), Latvia (40.3%), Denmark (36.1%), and Austria (33.4%).

The world electricity generation is predicted to be 24,112 TWh and still dominated by fossil fuel of about 66.12% or 16,000 TWh in 2022 [19]. As seen in Fig. 1.6A, fossil fuels' use increases steadily, with coal being the largest electricity generation source. Renewable electricity generation is expected to contribute over 8000 TWh (about 34%) in 2022, equal to China, India, and Germany's total power consumption. Hydropower will remain the largest source of renewable electricity generation in the forecast,

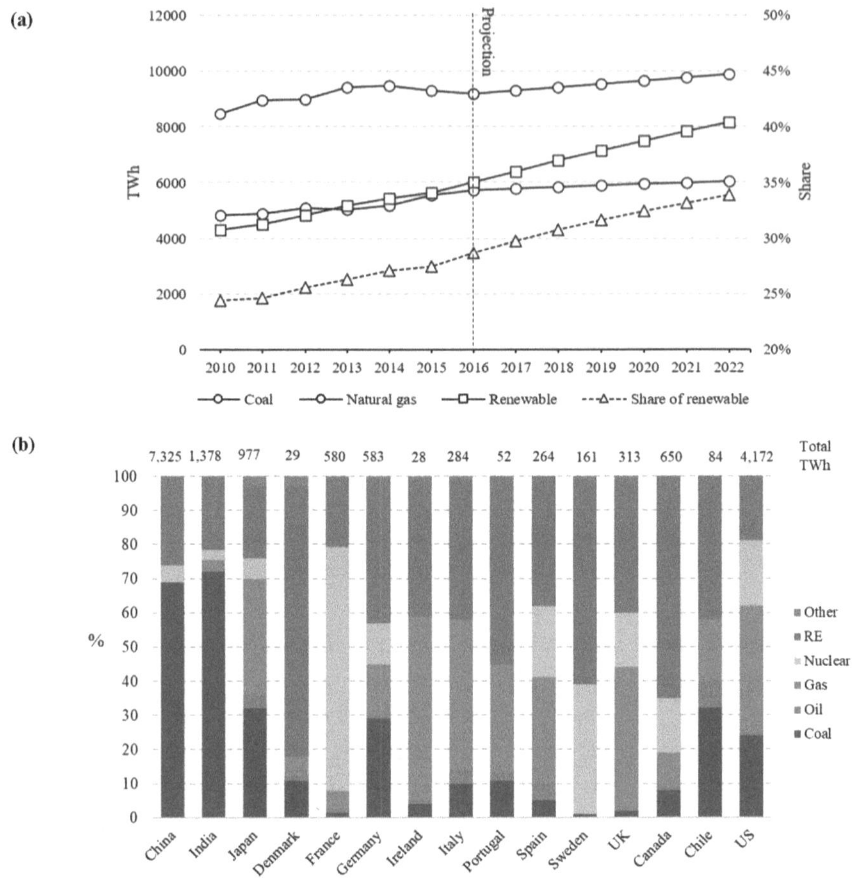

Figure 1.6 (A) Total electricity generation by fuel, 2010–22 and (B) electricity generation mix of selected countries in 2019.

followed by wind, solar PV, and bioenergy. The share of renewables in power generation has been increasing due to strong government policies in several countries to give attention to clean energy. The highest additional capacity (181 GW) of new renewable power plants was recorded in 2018 and now it accounts for around one-third of total installed power generation capacity worldwide. The fourth consecutive year of renewable power installations above 50% was recorded, including 64% in 2018. Fig. 1.6B shows the electricity generation mix of selected countries including Asia (China, Japan, India), Europe (Denmark, France, Germany, Ireland, Italy, Portugal, Spain, Sweden, United Kingdom), and America (United States, Canada, Chile) [20,21].

In 2020, the global health crisis due to the coronavirus COVID-19 pandemic also affected energy consumption, oil demand, and emissions level. It has slowed transportation, trade, and economic activities because of global lockdown measures. In the first quarter of 2020, global coal demand fell by almost 8% compared to the first quarter of 2019. Road transport and aviation activities have dropped to 50% and 60% of 2019, respectively, causing a complete halt in daily routines. This situation could also pose a significant threat to the timely deployment of renewables. The renewables momentum can also be disrupted without accurate responses. However, this uncertain situation is expected to be unsustainable in the long term and will not drastically change the global energy situation. Based on current policy scenarios, renewables' growth will continue, although not fast enough to offset the increase of energy demand due to population and economic growth. A supportive government is strongly required to promote an energy transition and determining the pace of deployment of renewables in the future.

1.3 Biomass waste as renewable energy

As one renewable energy source, biomass is considered a promising alternative fuel in the future. Biomass contains stored energy from the sun via the photosynthesis process. It is basically carbon neutral and evenly distributed widely across the globe. Biomass is categorized as a material with very complex components. This composition is a crucial parameter in determining the suitable conversion technology for each biomass. Although, in general, biomass mainly consists of cellulose, hemicellulose, and lignin, several extractives and minerals are also included, affecting material decomposition and its yields. Their conversion involves complicated reactions; therefore, it is complicated to clarify the mechanism and reactions during the conversion.

Although biomass can be classified in many ways, it can be divided into three main categories: natural crops (aquatic and terrestrial plantation), energy crops, and biomass waste. The latter is a by-product of biomass processing resulted from agricultural and industrial activities due to the high demand for food and other human needs. The massive increase in these industrial activities has led to a significant amount of biomass waste. Biomass waste utilization for energy purposes has some advantages: no conflict with other basic human needs for food or materials and environmental impacts reduction. The biomass waste is economically cheap; therefore, its effective utilization as an energy source is believed able to meet the financial requirement. It is also incoherent with the demand for industrial waste treatment to mitigate any further environmental impacts. The biomass waste itself can further be classified into human-made and natural biomass waste. The utilization of biomass waste can reduce the environmental problems caused by inappropriate treatment such as open burning or toxic waste disposal. The overview of biomass classification can be seen in Fig. 1.7.

Since animals eat plants or other animals, the stored chemical energy in animal waste can be classified as biomass. Animal waste generally contains organic matters, odors, bacteria, including nitrates. The chemical compositions vary depending on many factors such as feeding management, type of species, nutrient intake, etc. Traditionally, animal waste is utilized as organic manure to improve the soil nutrients for the plants. Currently, standard management is required to handle animal waste for promoting sustainability appropriately. Inappropriate treatment of animal waste can have environmental impacts such as water contamination or harming aquatic life. Via anaerobic digestion, animal waste can also be converted into biogas for energy purposes. The decomposition process is carried out by a group of bacteria to produce combustible methane and CO_2. Biogas projects are believed to generate a lot of interest in developing countries, especially in rural development. In terms of water content, animal waste can be classified into three types, solid, slurry, and wastewater. Different treatments are required depending on the purpose of animal waste utilization. Solid waste can directly be composted to produce fertilizer or dried up before being combusted to obtain heat energy. The slurry waste with

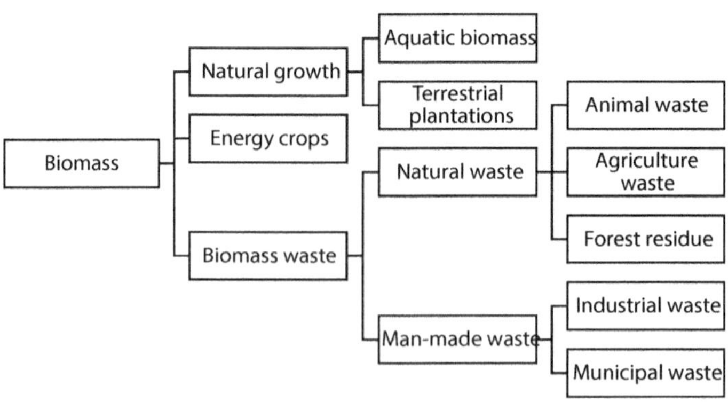

Figure 1.7 Biomass classification.

higher water content can be treated by composting for fertilizer production. The second option is methane fermentation for energy purposes. Lastly, the wastewater from animal waste can be treated to obtain either clean water or fertilizer [22].

Agricultural waste is defined as unwanted material from agriculture or agroindustrial activities such as stalks, leaves, roots, husks, bagasse, or straw. Cereal crops, mostly produced in low-income countries, are the major contributor to agricultural waste [23]. In reality, they are typically discarded to decompose naturally or burned. Due to the massive expansion of farming activities in developing countries, agricultural waste can cause environmental problems if they are not adequately managed, including the toxicity potential that can directly or indirectly affect human or animal life. The various utilization of agricultural waste has been applied, such as fertilizer, renewable energy sources (e.g., cooking, biogas), animal feeding, or other purposes. Large quantities of agricultural waste are promising potential as energy feedstock to obtain both economic value and environmental benefit. There are available routes to convert agricultural waste into useful energy; biochemical or thermochemical process. In the biochemical process, biogas and bioethanol can be produced using anaerobic digestion and fermentation, respectively. The thermochemical conversion via direct combustion, pyrolysis, gasification, or torrefaction is the most common route. These processes can produce heat, electricity, syngas, or biofuel. Depending on the product desired and the type of agricultural waste, the technological conversion decision can be different. The synergetic integration of agricultural process production and energy recovery of the waste is essential to enhance sustainable energy supply. Since the agricultural industry involves some of the most energy-intensive processes such as drying, the integration of thermochemical processes could realize a considerable economic and environmental benefit.

Forest waste or residue is an alternative feedstock for fuel or energy source. The waste is obtained from the harvesting process or silvicultural activity by controlling the forest's growth, composition/structure, and quality. Wood waste resulted from the wood processing industry usually has the highest percentage among the other forest residue. Industries that require wood supplies like sawmills, plywood, and panels can generate many by-products such as sawdust, off-cuts, trims, or shavings [24]. To date, forest waste is commonly processed to produce high-value wood pellets/chips as solid biofuel commodities. Furthermore, the main woody biomass conversion pathways are combustion, gasification, and pyrolysis.

Traditionally, combustion is performed to obtain heat energy via open burning or household stoves for cooking. It can also be integrated into industrial systems for heating, electricity generation, or other chemical production. Generally, the energy production cost will depend on many factors, including forest management, harvesting and transportation, conversion technology itself, and market situation.

Industrial waste derived from the primary biomass process can be recovered as an energy source. In a single industrial process, biomass waste can have high energy potentials such as residues in the pulp mill. The industrial waste can be either solid or liquid, depending on the processes involved in the industry. Combustion/cocombustion, pyrolysis, gasification, liquefaction, fermentation, and anaerobic digestion are the leading available energy conversion technologies. The proper utilization can significantly improve the economic benefit and minimize environmental impacts. The industry's appropriate waste recovery can also satisfy internal energy demand (heat or electricity) before the energy surplus is exported.

Municipal waste is one of the biomass energy sources that can be converted into electricity, fuels, or industrial chemical feedstocks. The amount of municipal waste is increasing due to human activities, resulting in environmental problems such as lack of landfill sites and contaminants' release. Mixed municipal waste can consist of plastics, rubber, paper, food waste, and yard trimming. The incinerator is the most well-known technique to reduce solid material (up to 95%) and recover energy from municipal waste by combustion process [25]. During the combustion process, the heat released is then utilized to produce steam for electricity generation (via steam turbine) or district heating. Since municipal waste typically has a high moisture content, the combustion temperature is relatively low, resulting in lower efficiency. Pyrolysis and gasification are attractive alternative processes to produce fuels from municipal waste over traditional combustion. Some valuable chemical commodities such as ethanol, synthetic gas can be recovered by series of these thermochemical processes. Biogas production is another option to utilize municipal waste via the biochemical pathway. Depending on many factors such as waste availability, waste composition, including the products desired, several conversion technologies can be adapted to convert municipal waste into more useful biofuels or electricity efficiently.

Since biomass waste is available at the boundaries of a mill or industrial activity, it can easily be processed as biomass fuel (e.g., POME in palm oil mill or black liquor (BL) in the pulp mill industry). As seen in Fig. 1.8,

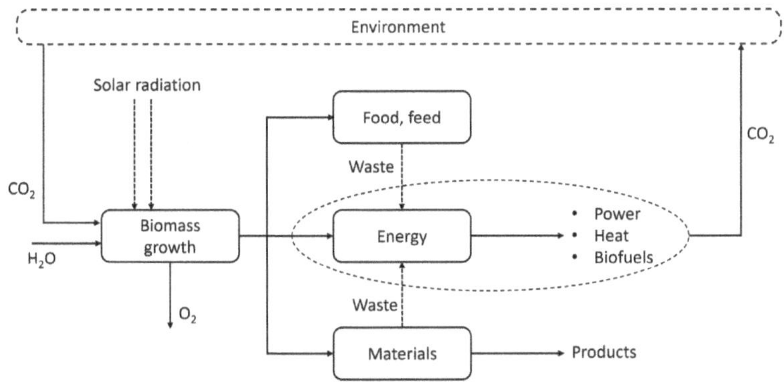

Figure 1.8 The cycle of biomass energy. *(Ptasinski K.J., Efficiency of biomass energy - an exergy approach to biofuels. Power Biorefin, 2017, https://doi.org/10.1002/cite.201770067.)*

biomass waste can be converted into electricity, heat, or biofuels (such as CH_4, H_2, bioethanol) via thermochemical or biochemical pathways. These waste biomasses may vary in terms of water content, chemical elements, particle size, and mass density. Their proper utilization can significantly improve the economic benefit and minimize the impacts on the environment. The appropriate waste recovery can also satisfy the internal energy demand before the surplus energy can be exported. The biomass waste utilization is also considered carbon-neutral due to its balance between carbon released during use and carbon absorbed when the plant grows. It is assumed that we do not consider the carbon emissions associated with harvesting and transporting the crops or industrial processing the biomass.

Among secondary energy forms, electricity, hydrogen (H_2), and liquid transportation fuels are energy carriers that can satisfy most energy needs in the future. They can be generated from various sources, including biomass waste, with different routes. By far, most of the electricity is produced by fossil fuel combustion in power plants. Electricity typically can be transmitted over a long distance from its generation plant to final users [26]. Consumers use electricity for daily activities such as lighting, cooling, heating, motion, refrigeration, and operating appliances. The development of H_2 has also attracted much attention from many researchers, including efficient production, storage, or distribution. As electricity, H_2 can be transported over vast distances via pipelines. It can also be effectively stored in gaseous, liquid, or metal hydrides form depending on end users. It has lower environmental impacts at the point of use [27]. Due to its high energy density and overall cleanliness, H_2 is considered a promising solution for challenges related to CO_2 emissions caused by fossil fuel usage, guaranteeing sustainable energy for the global population. However, hydrogen faces several major challenges, including its still-high production cost, inefficient storage and transportation, and low social acceptance. Hence, efficient production and storage systems for hydrogen are urgently needed, especially in terms of both technology and economy [28]. Hydrogen production can be performed in several ways, including the separation from hydrocarbons (reforming), water splitting (electrolysis), and biological production.

Furthermore, hydrogen is still dominantly produced from fossil fuels (about 95% of the total global production), including natural gas (nearly half of the total), coal (18%) and oil (30%) [29,30]. Electrolysis and biomass-based conversion only share about 4% and 1%, respectively [31]. Although the hydrogen production from any renewable becomes the long-term projection in the future hydrogen-based energy system [32], unutilized energy sources such as biomass waste are still potential, especially for short and medium scenarios.

1.4 Biomass waste properties

Depending on the source, biomass waste mainly consists of cellulose followed by hemicellulose, lignin (aromatic polymer), and a small number of other extractives. Cellulosic biomass is generally not used as food ingredients or fermentation substrates. As an animal skeleton's function, cellulose is a very complex polymer of 6-carbon sugar molecules (glucose) linked together in the crystalline structure. Since it is resistant to hydrolysis, pretreatments are usually required to break complex cellulose into simple sugars before being fermented to produce biofuel. Hemicellulose that is weaker than cellulose is complex polysaccharides made from various five- and six-carbon sugars. While cellulose is crystalline and strong, hemicellulose can easily be hydrolyzed into simple sugars. The lignin of cellulosic biomass is a class of complex organic polymers, which provides mechanical strength to the cell wall. During biofuel production, lignin is either left as a residue or recovered as an energy source via a thermochemical conversion process. The estimation of the chemical composition of selected biomass waste can be seen in Table 1.2. In some industrial waste, the chemical composition is degraded via complex chemical reactions during the industrial process. In this case, both raw biomass material and process conditions will significantly influence biomass waste feedstock. Some variables

Table 1.2 Chemical composition of selected biomass waste.

Selected biomass waste		Cellulose (%)	Hemicellulose (%)	Lignin (%)
Animal waste	Cow manure [33]	26.59	11.27	11.24
	Horse manure [34]	33.7	22.1	16.9
Agricultural waste	Leaves [35]	15–20	80–85	–
	Corn cobs [36]	33.70–41.20	31.9–36.0	6.1–15.9
	Rice husk [37]	33	26	7
Forest residue	Hard wood [38]	45	30	21
	Soft wood [38]	45	25	25
	Sawdust [39]	41.58	32.81	33.56
Industrial waste	Black liquor	–	–	25–31
	Empty fruit bunch [40]	24–65	21–34	14–31
	Palm kernel shell [40]	7	26	50
Municipal waste	Organic residue [41]	21–64	5–22	3–28

such as moisture content, proximate and ultimate analysis, chemical composition, and calorific value of biomass feedstock will determine the most suitable technology conversion, including the pretreatment process [1].

Generally, thermochemical conversion typically requires low moisture content (<50%), while the biochemical pathway can utilize feedstocks with high moisture content [42]. Although raw biomass's water content is in the range of 5%–35%, it can slightly be reduced by either natural or thermal drying. The biomass waste can also have high moisture content (>40%) due to water involvement, such as in the industrial process or animal farms.

Calorific value is the energy contained in fuel or feedstock, equivalent to the heat generated by its complete combustion. For biomass waste, the calorific value is commonly expressed in MJ kg^{-1}. Complete biomass combustion means that all the carbon, hydrogen, and sulfur is burned to carbon dioxide, water, and sulfur dioxide, respectively. There are two conventions in engineering practices to quantify the amount of heat produced in combustion, namely higher heating value (HHV) and lower heating value (LHV). In the HHV calculation, the latent heat of evaporated water during fuel combustion is also included. Condensing this evaporated water releases a lot of energy, and more heat practically can be recovered. Conversely, the LHV calculation does not include the heat released in the condensing water. It is assumed that the combustor has no secondary condenser; therefore, the produced vapor during fuel combustion is not recovered and released into the environment. We can conclude, the value of HHV is always higher than or equal to the LHV. The correlation between LHV and HHV can be written,

$$LHV = HHV - h_{vap}\frac{m_{water}}{m_{fuel}}$$

where h_{vap}, m_{water}, and m_{fuel} are (latent) heat of vaporization at the specified temperature, the mass of water, and the fuel mass, respectively. The HHV for solid fuel, such as biomass feedstock, can be obtained experimentally by employing a bomb calorimeter. A measurement using a bomb calorimeter requires an insulated and adiabatic space to perform complete fuel combustion. This approach is considered to be sophisticated, time-consuming, and costly [43]. General correlations have been developed to estimate heating values of fuel using ultimate or proximate analysis. The former measure the weight percent of carbon, hydrogen, nitrogen, sulfur, and ash from a fuel sample. However, the investigation of the elemental compositions via ultimate analysis is costly and requires sensitive equipment. The proximate analysis got more popular to determine the amount of fixed carbon, volatile matter, ash, and moisture content because of it being cheaper and more simple [44].

1.5 Biomass waste potential

Biomass always plays a vital role in the history of humankind. Before the fossil fuel era, biomass had been used for thousands of years as the primary energy source. In the early 20th century, biomass has become popular again after Henry Ford fueled his car using liquid biofuel or ethanol. In the 1970s, due to geopolitical conflict, the Organization of Petroleum Exporting Countries (OPEC) proclaimed an oil embargo, causing an energy crisis. This situation forced researchers and governments to discuss climate change and reduce fossil fuel use [45]. Among their exploration was a reassessment of fuel production from organic matter. Therefore, the interest in bioenergy is increasing due to concerns about energy security and growing environmental concerns. The bioenergy deployment is also seen as a potential stimulus for economic growth, especially in developing countries.

Globally, the available land is estimated to be 13 billion hectares (ha), of which 4.5 billion ha are suitable for crop production (see Fig. 1.9) [46]. Out

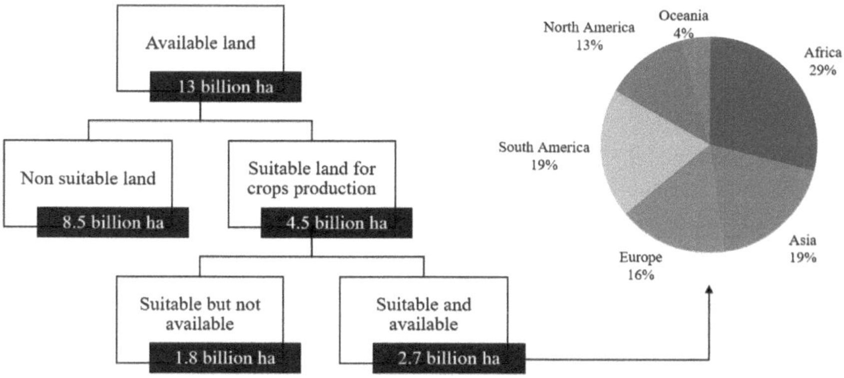

Figure 1.9 Share of suitable land available for crop production (total 2.7 billion ha).

of this 4.5 billion ha, the total amount of suitable land available for crop production is estimated at approximately 2.7 billion ha, due to several circumstances such as urban and protected areas. Among the suitable and available land for crops production, the African continent has the highest share of 29% or about 806 million ha, followed by South America (540 million ha), Asia (529 million ha), Europe (448 million ha), North America (352 million ha), and Oceania (113 million ha).

Biomass accounted for about 10% of global energy supply in 2017, or 55.56 EJ—a drop of 0.9 EJ over the previous year [47]. Although biomass has the largest share among other renewables (responsible for 70%), it is dominated by traditional biomass use as a fuel in developing countries for cooking and heat sources [48]. According to IRENA and World Bioenergy Association (WBA), the total biomass supply by 2030 could range 97−147 EJ/year and 130−154 EJ/year, respectively (see Table 1.3). IRENA estimates about 33−39 EJ/year of the total supply coming from energy crops. The WBA expects a lower potential for energy crops of about 18 EJ/year due to the difference in land availability assumed (230 million hectares vs. up to 900 million hectares). The potentials of forestry products according to the WBA is 70 EJ/year, higher than IRENA's range of 27−43 EJ/year.

It is also essential to look at local biomass potential, including its bioenergy market. In Southeast Asia, for example, the waste from the palm oil industry is increasing following the massive expansion. It is also indicated by the annual growth rate of 7.2% between the years 2016−21 based on the production forecast [50]. Consequently, high-potential biomass wastes from this industry, such as an empty fruit bunch (EFB), must be utilized with the

Table 1.3 Global biomass potential (EJ) by category for the year 2030.

Biomass category	IRENA [48]	WBA [49]
Energy crops	33–39	18
Residues and waste		
• Forest residues	27–43	70
• Agriculture and industrial waste	19–48	42–69
• Animal and municipal waste	18	
Total	97–148	130–154

appropriate technology to optimize its economic benefit and reduce the environmental impacts. Improper disposal practices, such as open burning, will cause many environmental problems. According to data from the Indonesian Ministry of Agriculture, palm trees plantations' total area was around $8 \times 104\ km^2$ in 2015. It was twice as much as in 2000 ($4 \times 10^4\ km^2$). This number is projected to increase to $1.3 \times 10^5\ km^2$ by 2020. The annual production of crude palm oil in Indonesia was 27.78 Mt in 2013. This product is expected to reach 37 Mt in 2019, with a yearly growth of 4.59%. Palm oil production is mainly located in Sumatera (70%) and Kalimantan (30%) islands.

Globally, the pulp and paper industry currently process about 170 Mt of BL per year (measured as dry solids), making BL a very significant biomass source [51]. BL comprises viscous alkaline solutions and has various compositions depending on the properties of wood and cooking conditions. Generally, it contains organic compounds of cellulose, hemicellulose, and lignin, in addition to alkaline salts such as sodium hydroxide and sodium sulfide [52]. A pulp mill generates 1.7–1.8 t-BL/t-pulp (dry content). BL thus represents a potential energy source of 250–500 MW/mill. Energy recovery from BL is sufficient to provide the electricity and steam required by a pulping plant. A pulp mill consumes about 700 kWh-electricity/t-pulp [53].

About 90% of the global rice production comes from Asian countries, and Southeast Asia provides 29% of it. It was assumed that, in Southeast Asia, over 100,000 rice mills have a potential of 16,720 MW generable power from 19 million tons of rice husks produced every year [54]. The high rice production has led to the production of a huge amount of agricultural wastes consisting of straw and husk. It leads to many problems associated with the improper disposal practices of the rice waste products, such as burning in the open fields during the peak harvesting season, resulting in pollution problems [55]. Therefore, advanced utilization of both rice husk and straw, including energy harvesting, is urgently demanded from both economic and environmental points of view.

1.5.1 Biomass waste management for bioenergy

Improper biomass waste treatments have caused many environmental problems and posed risks to human life, including land disposal and open burning. The scale of impacts on the environment also depends on the type of biomass and the extent of use [23]. Open burning practices are expected to contribute about 18% of total CO_2 emissions. Most residues are not utilized for energy sources because they are difficult to collect or used for other purposes. Therefore, mapping the biomass potential and determining the stage at which the biofuel is produced can also minimize the distribution cost and the risks associated with the biomass utilization (see Fig. 1.10). The use of waste is usually preferable in terms of economic and environmental benefits. On the other hand, if an available crop is grown solely for energy purposes such as in surplus land, biofuel production could lead to deforestation with a higher risk to the environment. For example, in a single biomass-based industrial process such as palm oil production, the by-product (e.g., empty fruit bunch) or waste (POME) can have high energy potential with lesser risks during energy recovery. However, the

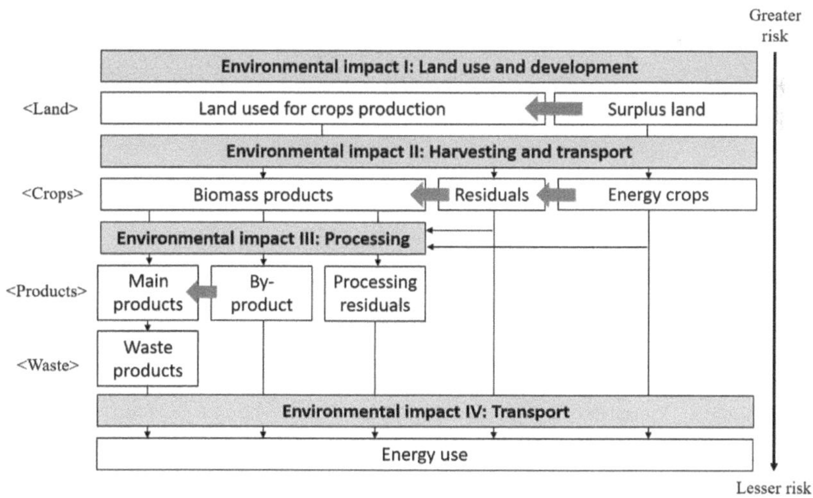

Figure 1.10 Stages of biomass utilization; impact assessment. *(Adapted from Investopedia, 2019. The World's top oil producers of 2019. Investopedia. https://www.investopedia.com/investing/worlds-top-oil-producers/.)*

recovery process involving waste cleaning needs to be verified due to the generation of concentrated organic wastewater. Therefore, assessments should be carefully performed at the processing stage. Furthermore, appropriate energy recovery (heat, electricity, or biofuel) can also satisfy internal demand before the energy surplus can be exported.

About one-third of global biodiesel supplies are produced in Europe following a ban on field burning, followed by Latin America (Brazil and Argentina)—contributing for a quarter. The produced biodiesel is processed from various biomasses. Furthermore, Brazil (48% of global exports), the United States (6%), and France (6%) are leading exporters of bioethanol, accounting for more than half of the global production [46]. In the United States, the agricultural residues (140–350 Mt) are used by industry as an energy source. In Europe, straw is the main waste from agricultural activities and commonly used for district heating, industrial processing, and domestic heating.

The use of biomass for power generation is widely reported. EU countries, the United States, and Japan dominate the current utilization for power generation. In 2012, the combination of municipal waste and BL from pulp mill accounted for nearly 80% of total installed capacity, followed by biogas (~20%) [46]. The increase of biomass utilization for power generation is usually supported by firm policy from the government. Electricity generated from biomass and waste in the United States totaled 70.6 TWh in 2018, utilizing organic feedstocks including wood and wood waste solids (30%), BL (28%), municipal solid waste (20%), and landfill gas (16%) [56]. Japan's electricity generation from biomass accounted for 1.6% of total generated power or 16.8 TWh in 2016. They also aimed to double biomass power generation to 32.8 TWh in 2030.

References

[1] Ptasinski KJ. Efficiency of biomass energy - an exergy approach to biofuels. Power Biorefin 2017;89. https://doi.org/10.1002/cite.201770067.

[2] Ng DKS, Tan RR, Foo DCY, El-Halwagi MM. Process development and resource conservation for biomass conversion. Chichester, UK: John Wiley & Sons, Ltd; 2015. https://doi.org/10.1002/9781118699140.

[3] Fujimoto S, Yanagida T, Nakaiwa M, Tatsumi H, Minowa T. Pinch analysis for bioethanol production process from lignocellulosic biomass. Appl Therm Eng 2011;31:3332–6. https://doi.org/10.1016/j.applthermaleng.2011.06.013.

[4] Inayat A, Raza M, Khan Z, Ghenai C, Aslam M, Shahbaz M, et al. Flowsheet modeling and simulation of biomass steam gasification for hydrogen production. Chem Eng Technol 2020;43:649–60. https://doi.org/10.1002/ceat.201900490.

[5] Jia Z, Chi R, Sun L, Tan H. Biodiesel processes energy improvement based on pinch and exergy analysis. Chem Eng Trans 2017;61:487–92. https://doi.org/10.3303/CET1761079.
[6] Heyne S, Thunman H, Harvey S. Extending existing combined heat and power plants for synthetic natural gas production. Int J Energy Res 2012;36:670–81. https://doi.org/10.1002/er.1828.
[7] Worldometer. World population clock: 7.8 billion people (2020). Worldometer; 2020. https://www.worldometers.info/world-population/. [Accessed 1 July 2020].
[8] Anderson MG, Katz PB. Strategic sourcing. Int J Logist Manag 1998;9:1–13. https://doi.org/10.1108/09574099810805708.
[9] Ferrini F, Bussotti F, Tattini M, Fini A. Trees in the urban environment: response mechanisms and benefits for the ecosystem should guide plant selection for future plantings. Agrochimica 2014;58:234–46. https://doi.org/10.12871/0021857201432.
[10] Investopedia. The World's top oil producers of 2019. Investopedia; 2019. https://www.investopedia.com/investing/worlds-top-oil-producers/. [Accessed 26 June 2020].
[11] IEA. Coal information 2019 analysis. Int Energy Agency; 2019. p. 22. https://doi.org/10.1787/coal-2018-en.
[12] Global Energy Statistical Yearbook 2018. Coal and lignite production data: world coal production. Enerdata. n.d. https://yearbook.enerdata.net/coal-lignite/coal-production-data.html. Accessed June 28, 2020.
[13] Global natural gas production: world gas natural statistics. Enerdata; 2018. https://yearbook.enerdata.net/natural-gas/world-natural-gas-production-statistics.html. [Accessed 28 June 2020].
[14] IEA. World energy balances 2019 – analysis. IEA; 2019. p. 23. https://www.iea.org/reports/world-energy-balances-2019. [Accessed 17 June 2020].
[15] REN21. Renewables 2019 global status report. 2019.
[16] Here's a list of countries committed to a net-zero emissions goal. World Economic Forum. n.d. https://www.weforum.org/agenda/2019/07/the-growing-list-of-countries-committing-to-a-net-zero-emissions-goal/. Accessed June 16, 2020.
[17] New Zealand commits to carbon neutrality by 2050 – with one big exception: NPR. n.d. https://www.npr.org/2019/11/07/777259573/new-zealand-commits-to-being-carbon-neutral-by-2050-with-a-big-loophole. Accessed June 16, 2020.
[18] Sweden had by far the highest share, lowest share in the Netherlands – the Netherlands and France: furthest away from their goals. n.d.
[19] Electricity generation by fuel, 2002–2022 – Charts – Data & Statistics. IEA. n.d. https://www.iea.org/data-and-statistics/charts/electricity-generation-by-fuel-2002-2022. Accessed June 17, 2020.
[20] Renewable energy institute (Japan). n.d.
[21] IEA. Monthly electricity statistics archives. 2012.
[22] Haga K. Animal waste problems and their solution from the technological point of view in Japan. Jpn Agric Res Q 1998;32:203–10.
[23] Tripathi N, Hills CD, Singh RS, Atkinson CJ. Biomass waste utilisation in low-carbon products: harnessing a major potential resource. Npj Clim Atmos Sci 2019;2. https://doi.org/10.1038/s41612-019-0093-5.
[24] Belyakov N. Sustainable power generation: current status, future challenges, and perspectives. Elsevier; 2019. https://doi.org/10.1016/C2018-0-01215-3.
[25] Cheremisinoff NP, Rosenfeld P. Handbook of pollution prevention and cleaner production. Best practices in the wood and paper industries, vol. 2. Elsevier Inc.; 2010. https://doi.org/10.1016/C2009-0-20361-8.
[26] Nejat Veziroglu T, Sherif SA, Barbir F. Hydrogen energy solutions. Environ Solut 2005:143–80. https://doi.org/10.1016/B978-012088441-4/50008-3.

[27] Aziz M. Integrated hydrogen production and power generation from microalgae. Int J Hydrogen Energy 2016;41:104–12. https://doi.org/10.1016/J.IJHYDENE.2015.10.115.
[28] Orhan MF, Dincer I, Rosen MA, Kanoglu M. Integrated hydrogen production options based on renewable and nuclear energy sources. Renew Sustain Energy Rev 2012;16:6059–82. https://doi.org/10.1016/j.rser.2012.06.008.
[29] Venkata Mohan S, Vijaya Bhaskar Y, Sarma PN. Biohydrogen production from chemical wastewater treatment in biofilm configured reactor operated in periodic discontinuous batch mode by selectively enriched anaerobic mixed consortia. Water Res 2007;41:2652–64. https://doi.org/10.1016/j.watres.2007.02.015.
[30] Dincer I. Green methods for hydrogen production. Int J Hydrogen Energy 2012;37:1954–71. https://doi.org/10.1016/j.ijhydene.2011.03.173.
[31] Das D, Khanna N, Nejat Veziroğlu T. Recent developments in biological hydrogen production processes. Chem Ind Chem Eng Q 2008;14:57–67. https://doi.org/10.2298/CICEQ0802057D.
[32] Cherry RS. A hydrogen utopia? Int J Hydrogen Energy 2004;29:125–9. https://doi.org/10.1016/S0360-3199(03)00121-6.
[33] Kumar P, Barrett DM, Delwiche MJ, Stroeve P. Methods for pre-treatment of lignocellulosic biomass for efficient hydrolysis and biofuel production. Ind Eng Chem Res 2009;48(8):3713–29. https://doi.org/10.1021/ie801542g.
[34] Zhang J, Loh KC, Lee J, Wang CH, Dai Y, Wah Tong Y. Three-stage anaerobic co-digestion of food waste and horse manure. Sci Rep 2017;7:1–10. https://doi.org/10.1038/s41598-017-01408-w.
[35] Prasad S, Singh A, Joshi HC. Ethanol as an alternative fuel from agricultural, industrial and urban residues. Resour Conserv Recycl 2007;50:1–39. https://doi.org/10.1016/j.resconrec.2006.05.007.
[36] Isikgor FH, Becer CR. Lignocellulosic biomass: a sustainable platform for the production of bio-based chemicals and polymers. Polym Chem 2015;6:4497–559. https://doi.org/10.1039/c5py00263j.
[37] Jackson MG. Review article: the alkali treatment of straws. Anim Feed Sci Technol 1977;2:105–30. https://doi.org/10.1016/0377-8401(77)90013-X.
[38] Hu G, Heitmann JA, Rojas OJ. Feedstock pre-treatment strategies for producing ethanol from wood, bark, and forest residues. BioResources 2008;3:270–94. https://doi.org/10.15376/biores.3.1.270-294.
[39] Ahmad ZS. Characterization of meranti wood sawdust and removal of lignin content using pre-treatment process. Natl Conf Postgrad Res 2016:598–606.
[40] Chang SH. An overview of empty fruit bunch from oil palm as feedstock for bio-oil production. Biomass Bioenergy 2014;62:174–81. https://doi.org/10.1016/j.biombioe.2014.01.002.
[41] Mcginnis GD, Wilson WW, Mullen CE. Biomass pre-treatment with water and high-pressure oxygen. The wet-oxidation process. Ind Eng Chem Prod Res Dev 1983;22:352–7. https://doi.org/10.1021/i300010a036.
[42] McKendry P. Energy production from biomass (part 1): overview of biomass. Bioresour Technol 2002;83:37–46. https://doi.org/10.1016/S0960-8524(01)00118-3.
[43] Majumder AK, Jain R, Banerjee P, Barnwal JP. Development of a new proximate analysis based correlation to predict calorific value of coal. Fuel 2008;87:3077–81. https://doi.org/10.1016/j.fuel.2008.04.008.
[44] Nhuchhen DR, Abdul Salam P. Estimation of higher heating value of biomass from proximate analysis: a new approach. Fuel 2012;99:55–63. https://doi.org/10.1016/j.fuel.2012.04.015.
[45] Mohr SH, Wang J, Ellem G, Ward J, Giurco D. Projection of world fossil fuels by country. Fuel 2015;141:120–35. https://doi.org/10.1016/j.fuel.2014.10.030.

[46] International Renewable Energy Agency. International renewable energy agency IRENA global bioenergy supply and demand projections a working paper for REmap 2030. 2014.
[47] Global bioenergy statistics 2019. World Bioenergy Association; 2019. p. 58.
[48] International Renewable Energy Agency. Global renewables outlook: energy transformation 2050. 2020.
[49] Kopetz H. Renewable resources: build a biomass energy market. Nature 2013;494:29–31. https://doi.org/10.1038/494029a.
[50] Zion Market Research. Palm oil market analysis by derivative (CPO, PKO, PKC, and others), and for edible oil, cosmetics, bio-diesel, lubricants, surfactants and other applications – global industry perspective, comprehensive analysis, and forecast 2015–2021. Maharashtra. 2016.
[51] Zhang Y, Yao M, Gao S, Sun G, Xu G. Reactivity and kinetics for steam gasification of petroleum coke blended with black liquor in a micro fluidized bed. Appl Energy 2015;160:820–8. https://doi.org/10.1016/j.apenergy.2015.01.009.
[52] Ji X, Lundgren J, Wang C, Dahl J, Grip CE. Simulation and energy optimization of a pulp and paper mill - evaporation plant and digester. Appl Energy 2012;97:30–7. https://doi.org/10.1016/j.apenergy.2012.01.014.
[53] Bajpai P. Pulp and paper industry, vol. 1; 2016. https://doi.org/10.1017/CBO9781107415324.004.
[54] Shafie SM, Shafie SM, Shafie SM. A review on paddy residue based power generation: energy, environment and economic perspective. Renew Sustain Energy Rev 2016;59:1089–100. https://doi.org/10.1016/j.rser.2016.01.038.
[55] Soam S, Kapoor M, Kumar R, Borjesson P, Gupta RP, Tuli DK. Global warming potential and energy analysis of second generation ethanol production from rice straw in India. Appl Energy 2016;184:353–64. https://doi.org/10.1016/j.apenergy.2016.10.034.
[56] Energy& environment specialist. London: JETRO; 2015.

CHAPTER 2

Process and products of biomass conversion technology

Arif Darmawan[1] and Muhammad Aziz[2]
[1]Agency for the Assessment and Application of Technology (BPPT), Puspiptek Serpong, Tangerang Selatan, Indonesia; [2]Institute of Industrial Science, The University of Tokyo, Meguro-ku, Tokyo, Japan

The development of biomass-to-energy conversion methods has been expanding significantly. Biomass conversion into useful thermal energy/heat, power, or fuels has been investigated across the world. The choice of conversion technology is affected by factors such as type and quality of biomass, final energy form, environmental standards, economic conditions, and project-specific factors [1]. Initially, the collected biomass (depending on the conditions) waste is upgraded by an appropriate method. It can include size reduction, densification, drying, and torrefaction [2]. The main biomass-to-energy conversion pathways can be classified into thermochemical and biochemical conversion, as shown in Fig. 2.1. Thermochemical conversion involves heat and chemical processes to obtain primary products from biomass waste. The main technologies of thermochemical conversion are combustion, gasification, pyrolysis, liquefaction, and thermochemical cycle. Biochemical conversion involves microorganisms to decompose and convert biomass into gaseous or liquid fuels. The most popular biochemical technologies are anaerobic digestion, fermentation, and photobiological hydrogen production. The primary products can be consumed directly or processed as intermediate fuels into final products. Furthermore, the final products can be utilized directly on-site, or it is stored through different types of methods and then transported and distributed to the consumers.

2.1 Biomass upgrading

The advantages of biomass as a chemical feedstock are the convenient chemical composition with relatively high carbon and hydrogen content [2]. However, some physical biomass properties can be a challenge for biomass conversions, such as low bulk and energy density and high moisture content. These problems should adequately be overcome with various processes, such as size reduction, drying, and thermochemical treatment, such as torrefaction. Direct use in a specific conversion process sometimes

Innovative Energy Conversion from Biomass Waste
ISBN 978-0-323-85477-1
https://doi.org/10.1016/B978-0-323-85477-1.00004-X

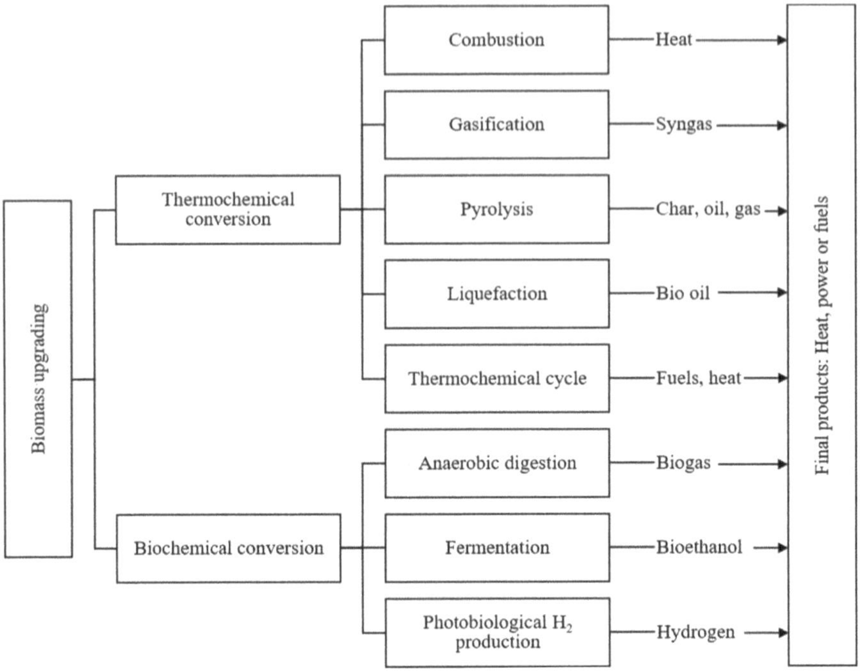

Figure 2.1 Main biomass-to-energy conversion pathways.

requires size reduction utilizing grinders, shredders, and chippers for finer feedstocks particles. Generally, most biomass has high moisture content. Drying of biomass or waste has been widely applied in industrial sectors such as food processing, pulp industry, and wastewater treatment. The energy required for this process also varies from 5% to 35%, depending on the type of industry. Drying is necessary to increase the calorific value, enhance the storability (more stable and longer storage and transportation), and improve the quality of the products.[3].

Drying is performed to remove water as vapor by transferring heat to wet solid material. The process itself depends on external conditions (temperature, humidity, and pressure) and dried solids' internal conditions. In the energy sector, biomass drying is usually required prior to the thermochemical conversion process (i.e., direct combustion, cocombustion with coal, pyrolysis, or gasification) to improve energy quality and reduce emissions (Table 2.1). In biomass combustion, boiler efficiency can be increased by 5%—10% if dried biomass is used instead of wet feedstocks [4]. The drying process of biomass commonly adopts two techniques, namely

Table 2.1 Several cases of biomass drying for energy purpose.

Feedstock	Type of dryer	Temperature, solids (%)	Conversion technology
Black liquor [7]	Steam tube rotary evaporator	115°C, 72.5%	Gasification
Empty fruit bunch [8]	Fluidized bed	142°C, 95%	Gasification
Algae [9]	Rotary steam tube dryer	106–194°C, 65%–97.5%	Steam gasification
Olive pomace	Direct-contact thermal screw dryer	180°C, 90%	Combustion
Black liquor [10]	Multi-stage dryer	80–133°C, 70%	Combustion

mechanical drying and thermal drying. The mechanical drying is utilized for feedstocks with higher moisture content (>50%) using equipment such as filter beds, pressers, and centrifuges [5]. In order to achieve higher solid concentration, thermal drying is performed using various conventional driers, including direct- and indirect-fired rotary dryer, conveyor dryer, cascade dryer, packed moving bed dryer, fluidized bed dryer, and flash or pneumatic dryer [4,6].

Torrefaction is conducted at a relatively higher temperature than the drying process (typically 200–300°C) to degrade the hemicellulose. Hence, other components that degrade at a higher temperature (such as lignin) might still exist in the products. Some investigations reported that increasing temperature resulted in biomass densification and decreased mass yield [11–13]. An experimental study was performed by Chen et al. [13] to understand the degradation of hemicellulose, cellulose, and lignin. The study was conducted at temperatures 230, 260, and 290°C. The torrefaction with 230°C has a small impact on the biomass sample by releasing only moisture and light volatiles. The mass losses of hemicellulose, cellulose, and lignin were 2.74%, 1.05%, and 1.45%, respectively. The torrefaction at higher temperatures (260°C) caused a certain amount of hemicellulose pyrolyzed of about 37.98%, whereas cellulose and lignin were only barely affected. Furthermore, when biomass underwent the torrefaction at 290°C, large amounts of hemicellulose and cellulose were destroyed with a mass loss of 58.33% and 44.82%, respectively.

2.2 Thermochemical conversion

2.2.1 Combustion

Combustion is the common and oldest thermochemical process to produce heat by reacting fuel with oxygen. Biomass combustion is a series of chemical reactions by which carbon and hydrogen are oxidized to carbon dioxide (CO_2) and water vapor, respectively. Therefore, biomass rich in carbon and hydrogen is preferable as feedstock for combustion. Oxygen, nitrogen, and a small amount of sulfur present in the biomass also react with air or oxygen. The maximum flame temperature will mainly be affected by the amount of air supplied and the biomass's heating value and moisture content. The moisture content of biomass is an essential factor during combustion. Biomass upgrading by drying is sometimes required before burning to increase both the boiler's performance and efficiency and lower air emissions. High water content can reduce combustion performance and produce undesired reaction yields [14]. The general combustion reaction for biomass can be expressed as follows:

$$\text{Biomass} + \text{oxygen} \rightarrow \text{Carbon dioxide} + \text{water vapor} + \text{heat}$$

Combustion of biomass in a furnace occurs in four main stages: drying, devolatilization and gas combustion, char combustion, and ash formation, as seen in Fig. 2.2. Initially, the remaining water content starts to evaporate when biomass is heated. Devolatilization is a decomposition process that commences at approximately 200–300°C. Temperature, pressure, residence time, particle size, and biomass type can influence the devolatilization rate [15]. During this stage, the hemicellulose decomposes and releases more volatile, less tar, and less char at lower temperatures. Lignin often yields more char during this process. Subsequently, the volatiles and tars released from a biomass material are combusted in the presence of oxygen. The volatile combustion is much faster than the remaining char. In the next stage, the char reacts with oxygen, forming mainly CO and CO_2 gases.

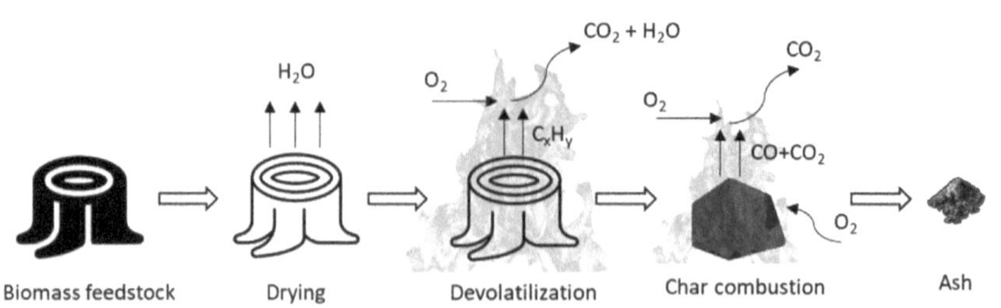

Figure 2.2 Stages in the combustion of a solid biomass particle.

In some cases, biomass combustion is performed in a furnace system designed for coal firing, with or without slight modification. For heat and power applications, there are three prominent types of combustion techniques: grate firing, fluidized bed, and pulverized fuel. These methods have the potential to release airborne pollutants [16]. Grate firing, the first combustion system used for solid fuels, is made to operate continuously by moving biomass material slowly through the boiler. The use of gate firing has some advantages, such as fuel flexible and low capital and operation costs. However, during the combustion, the fuel is not well mixed with long residence time. In this method, the air is supplied through holes in the grate. Grate firing can also be designed to remove ash into a container by a screw conveyor automatically. The fluidized bed, a newer combustion method, uses a medium like sand to the mixed particles and air effectively. Fuels supplied are reduced to 2–5 mm or larger by pelletizing or cutting into a chip to satisfy the fluidization process requirement. During the combustion process, the solid mixture is suspended by an upward movement of air. It results in proper mixing between solid materials and air.

Lower emission can also be achieved by controlling the flame temperature. Fluidized bed relatively has long residence time and higher cost. Furthermore, the pulverized fuel method is usually adopted for large-scale operation by grinding the feedstock to the powder before the combustion. It has a shorter residence time and better performance compared to the other techniques. Fuel particles usually are reduced to a wide range of 10–1000 μm or more [16]. High efficiency can also be achieved. Unfortunately, a higher temperature operation can trigger corrosion inside the furnace. The pulverized fuel can be utilized for cofiring with coal instead of dedicated biomass firing on a large scale to mitigate the problem related to the biomass's high chloride content [1] (Table 2.2).

The combustion is used for converting chemical energy stored in biomass into heat. The heat generated is then consumed directly or employed to produce steam for either heating or power generation. The hot steam can be used for heating in industrial activities, building, drying, and other operations requiring high temperatures. Biomass wastes such as agricultural residues, dried black liquor, municipal solid waste, and refuse-derived fuel (RDF) are commonly utilized by combustion in industrial activities. Energy recovered can be used for industrial operations that require high temperature or electricity supply. Therefore, a cogeneration system (or combined heat and power/CHP) is often adopted to increase overall efficiency by more than 90%. In a power plant, biomass combustion

Table 2.2 Comparison of the main biomass combustion technologies.

Technique of combustion	Advantages	Disadvantages
Grate firing	• Good fuel flexibility • Lower capital and operation cost	• poor mixing, • Long residence time • Low efficiency
Fluidized bed	• Good gas—solid contact and mixing • High efficiency • Lower NOx and SOx emission	• Long residence time • High capital and operation cost
Pulverized fuel	• Good operation performance • Short residence time • High efficiency	• Additional energy for grinding • Corrosion problem (at high temperature)

or cocombustion with coal is recovered by heat exchangers to generate steam to rotate a steam turbine. Due to slagging and fouling issues caused by high ash content, most biomass materials are cocombusted with coal to avoid technical problems. Retrofitting an existing coal-fired power plant to a biomass-coal cofiring system has a lower cost than building a new biomass power plant. Depending on plant size, quality, and share of biomass, coal-biomass cofiring power plants can have net efficiency of 36%—44% [17]. Since biomass has lower sulfur and nitrogen content than coal, the cofiring process can further reduce the SOx and NOx pollutants. Biomass cofiring with coal is also preferable on a large-scale power plant in terms of sustainable biomass supply. To date, a biomass share in continuous operation is usually less than 20%.

2.2.2 asification

Unlike combustion, in which fuel is fully oxidized, gasification is a partial oxidation process at a high-temperature range of 750—1100°C using gasifying agents such as air, steam, or oxygen [2]. Although gasification occurs with very complex reactions, the overall gasification processes for biomass ($C_nH_mO_k$) can be written as,

$$C_nH_mO_k + jO_{2(g)} + (sH_2O) \rightarrow \left(\frac{m}{2}+s\right)H_{2(g)} + (k+h+2j)CO_{(g)} + (n-k-s-2j)C_s$$

The gas produced usually is a mixture of syngas (H_2, CO, CH_4) and CO_2, H_2O, N_2, and light hydrocarbons. The syngas has been successfully applied for various purposes, including the generation of electricity and as an intermediate to produce H_2, dimethyl ether (DME), methanol, ethanol, or other fuels. The gasification process is usually combined with a power generation system, called integrated gasification combined cycle (IGCC), to improve the thermodynamic efficiency. The intermediate syngas has more flexibility to produce a wide range of fuels, which can replace petroleum-based refineries. Like combustion, gasification can also be used for any type of biomass feedstocks. One of the main issues of biomass gasification is tar and unwanted contaminants such as alkali compounds, resulting in a higher cost for gas cleaning [1]. The biomass gasification performance can be influenced by the type and design of the gasification reactor/gasifier, biomass characteristic, process temperature, catalyst, and gasifying agents. Gasification technologies are generally divided into three main types: fixed or moving bed, fluidized bed, and entrained flow, as seen in Fig. 2.3.

2.2.2.1 Moving bed gasifier

The moving bed is further subdivided into the updraft, downdraft, and crossdraft gasifier. These gasifiers usually operate at low gas velocity, high carbon conversion with long residence time [18]. They are also suitable for heat and electricity generation at a small scale [19,20]. The use of moving bed gasifiers for biomass fuels conversion has been widely reported, such as for mixtures of the chicken manure with wood [21], wood fuels [22–25], municipal solid wastes [26,27], and dairy biomass [28]. These studies also discussed the effect of various aspects such as biomass characteristics, gasifying agent, and reactor conditions on gasification performance.

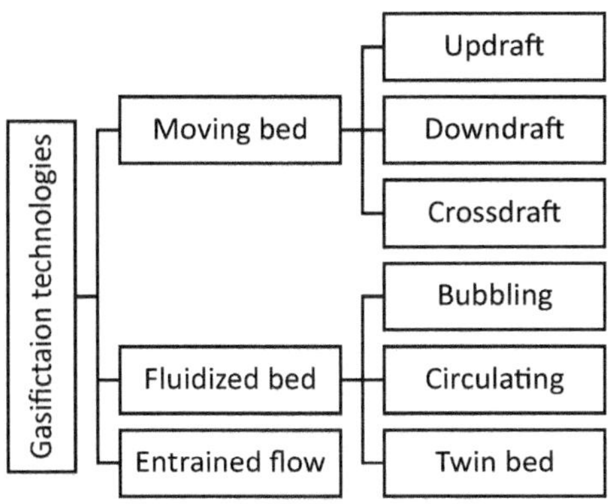

Figure 2.3 Types of gasifier.

Updraft gasifier is the simplest and commonly adopted for solid biomass. The solid biomass fuel is introduced from the top, and the air is supplied from the bottom of the gasifier, therefore flows in counter-current configuration (see Fig. 2.4A). The biomass moves downward, passing through the drying, devolatilization/pyrolysis, reduction/gasification, and combustion/oxidation zones. Updraft gasifier is easy to operate and has higher efficiency than combustion-based systems, although the reactor has slagging potential if it reaches a certain combustion temperature. To avoid slagging problems, operation temperature in the combustion zone should be carefully adjusted by controlling air/oxygen supply. The hot gas produced during the car combustion travels up and provides heat for the reduction process. Devolatilization occurs in lower temperatures and has tar (typically 10%–20% by weight of the fuel) at temperature range 200–500°C [29].

The downdraft or cocurrent gasifier is also commonly employed to convert solid fuels into gaseous products. In the downdraft gasifier, the solid biomass and air move cocurrently, and gas produced at the gasifier's bottom. This process has lower tar since the devolatilization product, tar, contacts oxygen in the combustion zone, as seen in Fig. 2.4B [29]. Downdraft gasifiers typically have a capacity of 10 kW–1 MW [30]. The biomass feedstocks with moisture content less than 30% are preferable for this type of gasifier. The higher water content will result in a lower quality of the produced gas [31]. Some modifications can be performed in the biomass feeding system, air supply system, discharge system, or recirculating system of producer gas to improve the gasifier performance.

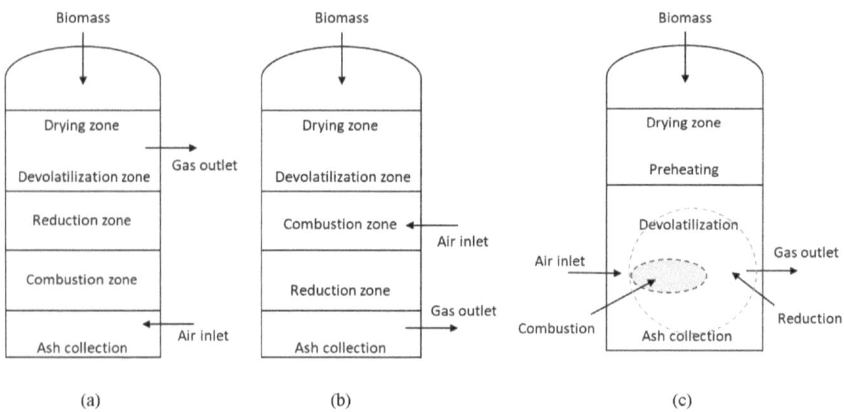

Figure 2.4 Schematic of an (A) updraft, (B) downdraft, (C) crossdraft biomass gasifier.

The crossdraft gasifier is also a cocurrent gasifier, which is solid fuel biomass introduced from the top, and the air is passed through a nozzle from the side of the gasifier [30]. From the configuration shown in Fig. 2.4C, the gas produced is released from its sidewall opposite to the air inlet. The produced gas leaves the gasifier at relatively high temperatures, about the same as gasification. Therefore, if no heat recovery, the gasifier's energy loss is increased, and its thermal efficiency is significantly reduced. The crossdraft is usually operated on a small scale of less than 10 kW and suitable for charcoal or pyrolyzed fuels with lower light particles. The summary of the comparison between the moving bed technologies is listed in Table 2.3. Moving bed gasifiers have taken continuous growth with its design, working biomass fuel, and performance parameters issues [32].

2.2.2.2 Fluidized bed gasifier

Fluidized bed reactors have some advantages such as excellent heat and mass transfer rates, scalability, good mixing capacities, and excellent reaction rates and conversions [33–35]. In this scheme, the biomass particles are moved freely throughout the bed by maintaining gas velocity injected from the

Table 2.3 Comparison of the moving bed gasifiers.

	Updraft	Downdraft	Crossdraft
Operation	The solid biomass fuel is introduced from the top and air is supplied from the bottom of the gasifier (counter-current configuration)	The solid biomass and air move cocurrently and gas produced at the bottom of the gasifier.	Solid fuel biomass is introduced from the top, and the air is passed through a nozzle from the side of the gasifier. The gas produced is released from its sidewall opposite to the air inlet.
Advantages	Suitable for fuel with high moisture, easy to operate	Lower tar and particulate in the gaseous product.	Simple design, good circulation in the combustion zone
Disadvantages	• High tar yield • Low heating value gas • Scale limitation	• Scale limitation • Moisture-sensitive • Low heating value	• More complicate to operate • Slagging issues • High carbon yields in the ash • Small scale

bottom or side part of the reactor. In fluidized bed gasification, air, steam, or oxygen can be used as a gasification agent. The bed material commonly uses sand, olivine, limestone, dolomite, or alumina, and the gasifiers operate below bed melting temperatures [33]. As the syngas leaves the fluidized bed reactor, a cyclone is used to separate the solid materials, including unreacted biomass, char, and the bed material, and return them to the fluid bed. Fluidized beds are suitable for particle sizes below 6 mm [36] and a wide range of moisture content [37,38], and preferable for the feedstocks with low ash and alkali metals contents.

Fluidized bed reactors can be mainly classified as bubbling, circulating, and dual fluidized bed. In a bubbling fluidized bed gasifier, the gasifying agent, such as air/oxygen or steam, is introduced from the reactor's bottom at a certain velocity (see Fig. 2.5A). The fluidization results in uniform temperature and good particle mixing across the bed. Silica or alumina materials are usually used due to high specific heat capacity and high operating temperatures. Catalysts also can be used as a fluidizing agent to increase the efficiency of biomass conversion. To avoid defluidization due to bed particle agglomeration, the use of biomass with high ash melting temperatures or operation at temperatures lower than ash melting temperature is advised [35,39]. Circulating fluidized beds requires a higher flow rate of the fluidizing agents up to three to five times more than the bubbling configuration, to move most of the solid and ungasified particles and recirculate them to the fluid bed. This configuration can increase the

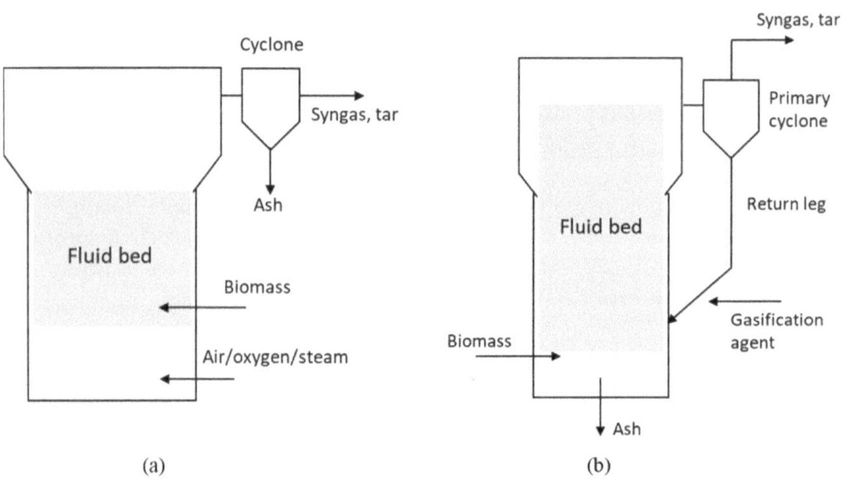

Figure 2.5 Schematic of (A) bubbling and (B) circulating fluidized bed gasifiers.

reaction rates, heat transfer, mass flow, carbon conversion efficiencies, and low tar production. The circulating fluidized bed gasifier is suitable for a large-scale process, up to 1 MW or higher. It also has flexibility in terms of fuel particle size and type. The disadvantages of the circulating fluidized-bed gasifier include (1) minimum transport velocity being influenced by the size of biomass fuels, (2) requirement of more complex design, (3) difficult control of the process, and (4) it being more expensive [40].

Dual or twin circulating fluidized beds consist of two chambers: a bubbling fluidized bed gasifier and a circulating fluidized bed combustor. The configuration of this gasifier can be seen in Fig. 2.6. Biomass feedstock and steam are supplied into the bubbling fluidized bed gasifier, enabling the gasification stage at bed temperatures of about 850–900°C. Subsequently, the tar produced and the solid materials such as char, residues produced in the bubbling bed, and part of the bed material are distributed to the circulating bed (combustor), where it is oxidized with air to generate heat and flue gas. The flue gas leaves the reactor via cyclone, and the hot bed

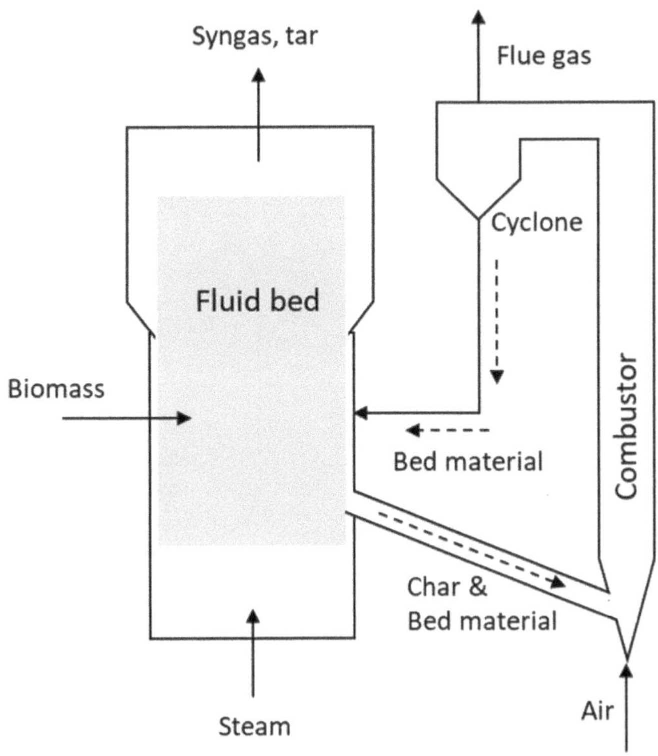

Figure 2.6 Schematic of dual fluidized bed reactors.

material returns to the bubbling bed chamber to provide heat for the endothermic steam gasification process. This mechanism can result in a lower tar level with moderate HHV of syngas produced. The dual bed gasifier has good scale-up potential and is applicable for a wide range of biomass fuels. It can be adopted for pine barks and sawdust, wood chips and pellets, grass, dried coffee grounds, cedar, oak sawdust, almond shells, municipal solid waste, and sewage sludge [41,42]. This technology has been successfully implemented at a demonstration scale, and design optimizations have been conducted to improve the gasification performance [43]. Among the drawbacks, dual fluidized gasifier has more complicated construction and operation.

2.2.2.3 Entrained flow gasifier

The last gasifier system to be described is the entrained-flow reactor. Entrained-flow gasifiers mainly have two types depending on how and where the fuel is fed into the reactor. Solid biomass particles can be fed from the top center with the oxidant (cocurrently) or the side of the gasifier counter-currently, as seen in Fig. 2.7A and B. In top-fed entrained-flow reactors, fine biomass particles (typically less than 100 μm in size) along

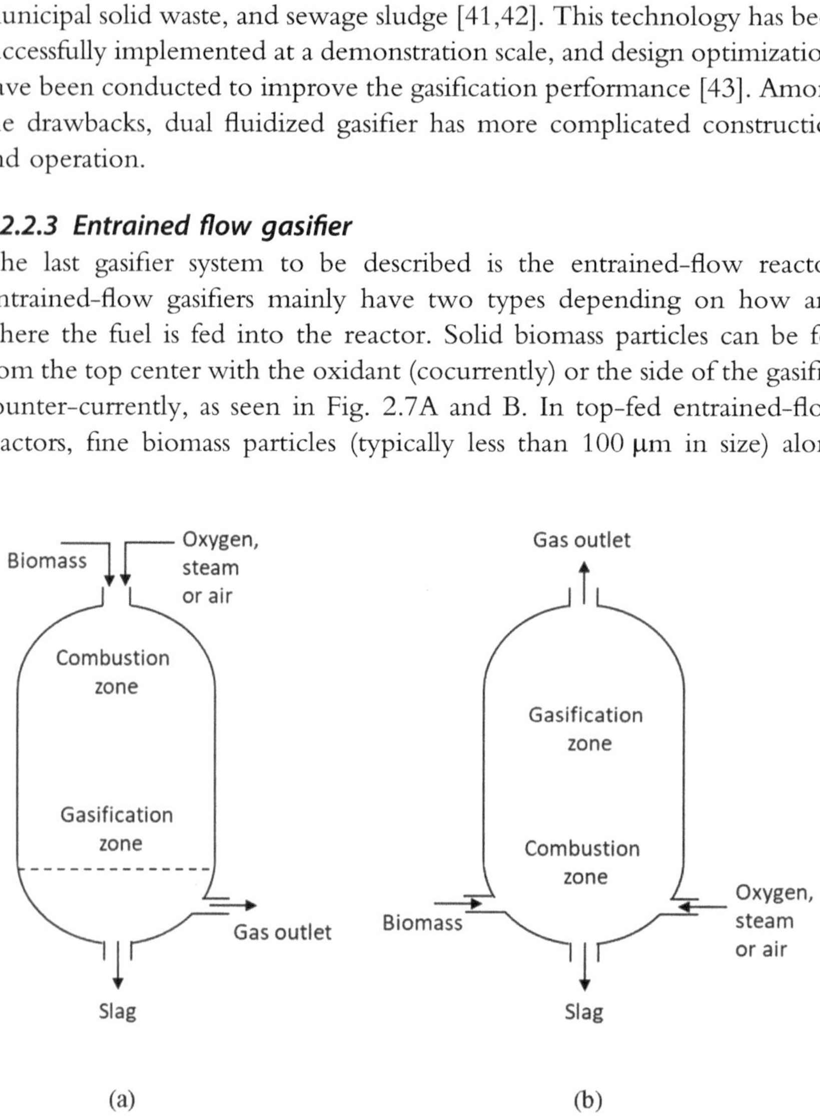

Figure 2.7 Schematic of two main types of entrained flow gasification reactors: (A) top-fed and (B) side-fed.

with oxygen and steam (air is rarely used) enter the gasifier cocurrently. The combustion reaction can take place right at the entry point of oxygen [44]. The gas velocity in the reactor is sufficiently high to entrain the powdered fuel particles fully. Slurry-fed gasifiers need additional reactor volume for evaporation of a large amount of water mixed with the fuel. The gasification reactions occur at a very high rate with high carbon conversion efficiencies (98%–99.5%). Entrained flow gasification typically operates at high temperatures around 1200–1500°C and 20–70 bar pressure, where powdered fuel is entrained in the gasifying medium. The properties of producer gas are mainly influenced by biomass fuel characteristics and gasifier operating conditions such as temperature, pressure, gasifying agent, biomass/oxygen ratio [45]. Entrained-flow gasifiers can handle practically any feedstock and produce clean, tar-free syngas. Since operated at high temperatures, the melting ash forms a slag.

2.2.2.4 *Entrained-flow gasifier*

The fine biomass fuels can be fed to the entrained-flow gasifier in either a dry or slurry form. The slurry feed has a simpler operation with higher operating pressure than dry feed. Due to this additional water, the syngas will contain a higher H_2 to CO ratio but lower cold gas efficiency. The water in the slurry feed is heated to the operating temperature using part of the feedstock, which has a penalty on the cold gas efficiency. Entrained-flow gasifiers typically exhibit the following characteristics [1,46]:

- Fuel flexibility of solid biomass feedstocks but also high efforts for fuel pretreatment
- Requires large oxidant, mostly using oxygen
- The uniform temperature inside the gasifier
- Slagging operation
- Short reactor residence time
- High carbon conversion, but low cold gas efficiency
- A high level of sensible heat in the product gas, heat recovery is required to improve efficiency
- Syngas consists of mainly H_2, CO, and CO_2 with very low tar and methane content.

2.2.3 Pyrolysis

Biomass pyrolysis is the thermal cracking of biomass fuels with the absence or very low concentrations of oxygen. The process typically takes place at temperature >300°C [47]. The main products of the pyrolysis process are liquid, gas, and char. Operating parameters and biomass properties determine the amount of each pyrolysis product. The product of pyrolysis can be further upgraded into fuel and chemicals via refining methods. Pyrolysis process employing biomass feedstocks has been widely reported, such as rice husk, sawdust pellets, cornstalk, and microalgae [48].

The maximum yield of an organic liquid (pyrolytic oil or bio-oil) from thermal decomposition may be increased to as high as 80% (dry weight) through proper choice of heating rate and pyrolysis temperature, and evacuation of product from the reaction zone [49].

The pyrolysis process can be divided into three types: flash, fast, and slow/conventional pyrolysis (see Table 2.4). Flash pyrolysis occurs at a higher temperature, around 700–1000°C. It is conducted at a very short residence time and a higher heating rate than the other types of pyrolysis. The main products of flash pyrolysis are mainly gases and liquid/bio-oil and the remaining biochar. Fast pyrolysis produces mainly bio-oil and occurs at a moderate temperature with a short residence of 0.5–2 s. This method has developed considerably in recent years to produce bio-oil and offered advantages in storage and transport. Compared to flash and slow pyrolysis, fast pyrolysis has higher energy efficiency and environmental acceptability. The fast pyrolysis method typically has bio-oil, gas, and char yields, 70–75, 13, and 12 wt% (expressed on dry biomass basis), respectively. Since the fast pyrolysis is an endothermic process, heat supply is required by complete or partial combustion of the produced gas or char. The bio-oil product is

Table 2.4 Classification of pyrolysis methods with differences in temperature, residence time, heating rate, and major products.

Flash pyrolysis	Fast pyrolysis	Slow pyrolysis

• temperature 700–1000°C
• short residence time <0.5 s
• high heating rate
  ◦ the main products are gases and bio-oil
	• medium temperature 400–600°C
	• short residence time 0.5–2 s
	• high heating rate
	  ◦ the main product is bio-oil. Gases and char are also produced
	  ◦ lower investment at small scale
		• at temperature >300°C (400–500°C),
		• long residence time 5–30 min.
		• lower heating rate
		  ◦ the products are gases, char, and bio-oil (tar)

potential as an intermediate for power generation or fuel production after upgrading and fractionation. The properties of bio-oils via a thermochemical process are very different from conventional petroleum-derived fuels [50]. Furthermore, a significant upgrading process is required before being used as transportation fuels, including economic feasibility assessment. To date, biomass fast pyrolysis plants have been used for industrial combustion purposes.

A schematic of fast pyrolysis system is shown in Fig. 2.8 for the noncatalytic process. Depending on biomass properties, biomass is initially pretreated by the drying process for feedstocks with high moisture content and grinding before entering the main pyrolysis reactor. Sand is usually employed as a heat carrier for the rapid heating of biomass. Subsequently, the hot vapors were distributed to a condenser and quickly condensed to reduce secondary reactions. The char and heat carriers are sent for combustion with air in a regenerator. The hot carrier is then recirculated back to the pyrolysis reactor. The gas producer can be combusted with char for drying and fluidization in the reactor. Fast pyrolysis can be conducted using various feedstocks, pyrolysis reactor designs, and configurations, including with the addition of catalysts and under a range of operating atmospheres [50–52].

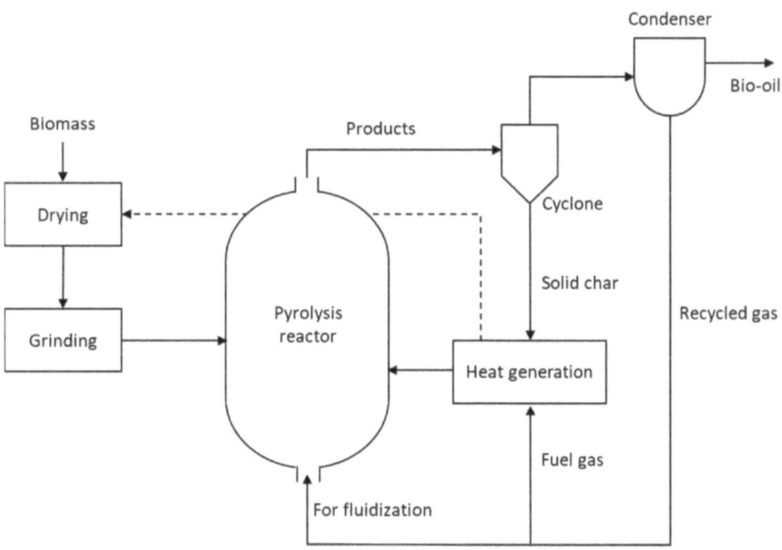

Figure 2.8 Schematic of a fast pyrolysis system for biomass.

2.2.4 Biomass liquefaction

As discussed earlier, the fast pyrolysis produces a relatively low quality of liquid biofuels and needs significant upgrading processes. Another promising method to produce renewable fuels, especially liquid biofuels, is biomass liquefaction. Liquefaction technologies are aimed to produce high-quality bio-oils that can also process high-moisture feedstock such as algae, organic food residues, bio-solids, and bio-sludge into transportation fuels. Liquefaction technologies can be performed via hydropyrolysis or hydrothermal liquefaction [53].

Hydropyrolysis is pyrolysis that is carried out under a hydrogen environment to increase process stability. Hydropyrolysis is usually performed using a catalyst and integrated with hydroconversion to produce gasoline and diesel range hydrocarbons. The integrated hydropyrolysis and hydroconversion consist mainly of hydropyrolysis, hydroconversion, upgrading system, and hydrogen production unit. This typical scheme has been developed by the US Department of Energy [54,55], as shown in Fig. 2.9. Biomass in the presence of hydrogen and catalyst is heated to 400−450°C in a fluidized bed reactor. Unlike in the fast pyrolysis, this hydropyrolysis process is exothermic; therefore, additional heat is not required. After temperature is reduced to <400°C, the bio-oil (vapors) products are distributed to a hydroconversion reactor. In this step, bio-oil is cracked in the presence of hydrogen and catalyst to produce a lighter oil fraction. The hydroconversion products are sent to a three-phase separator to separate them into vapor and two types of liquids: bio-oil and water. In the H_2 production unit, steam methane reforming (SMR) and pressure swing adsorption (PSA) are used to convert light gases into hydrogen that is sufficient to supply the integrated process. The gasoline and diesel are produced in distillation and upgrading units, and the remaining flue gas can be supplied into the hydrogen production unit.

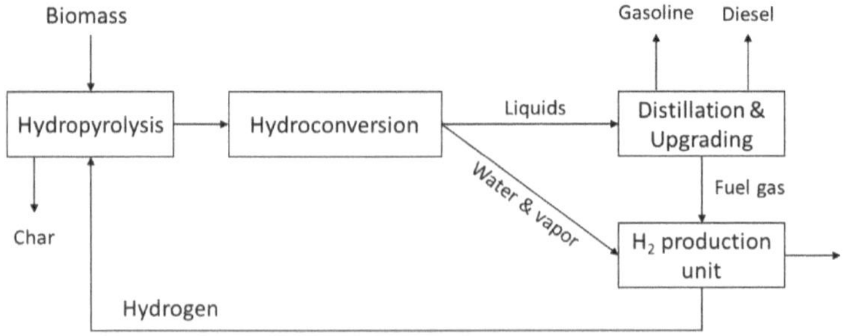

Figure 2.9 Schematic diagram of integrated hydropyrolysis and hydroconversion.

The second option of the liquefaction method is by adopting hydrothermal liquefaction (HTL) of biomass or wet pressurized pyrolysis. HTL is typically carried out at a moderate temperature of 250−450°C and high pressure of 10−35 MPa with the presence of water and catalyst. HTL is also an appropriate technology for bio-crude oil production from biomass with high moisture content such as black liquor, algae, food waste, municipal sludge [56,57]. Hence, there is no additional energy for the drying process. In the HTL process, water acts as a reactant to bring fast, homogeneous, and efficient reactions near the critical point. The catalysts are used to remove heteroatoms (oxygen, nitrogen, sulfur), enhance bio-oil (organic phase) yield, and enable better control over bio-oil composition [53]. The quality of HTL products is mainly influenced by the type of biomass, operating conditions such as residence time and temperature, and catalyst. Although HTL is a complex process, it can simply be divided into three main steps, depolymerization, decomposition, and recombination [58]. Briefly, the biomass feedstock is depolymerized and decomposed into small compounds. The decomposition of biomass monomers involves cleavage, dehydration, decarboxylation, and deamination. The highly reactive compounds are then polymerizing and forming bio-crude liquid, gas, and solid yields. HTL is considered a complex mechanism since biomass is a complex mixture containing carbohydrates, lignin, proteins, and lipids.

The bio-oil yields from HTL typically have an oxygen content of between 10 and 20 wt% and a heating value of between 30 and 35 MJ/kg, which is much better than the bio-oil product from the conventional pyrolysis. HTL is considered to have an advantage in terms of energy efficiency. It consumes 10%−15% of the biomass feedstock's energy content, resulting in an energy efficiency of about 85%−90% [59]. The reactor design for the HTL process can be either batch or continuous. The latter requires a feeding system that operates under pressure and includes either slurry-based pump or lock and hopper systems for large particles [59]. To date, HTL is still in a transient state from a lab−pilot scale to pilot−industrial scale, and no commercial-scale facility has been reported to be in operation. More investigation, including HTL mechanisms, is needed to develop large-scale systems. An operation of HTL at high pressure also requires proper materials to avoid corrosion and a reliable pumping system for water slurries/solid, especially in a continuous process.

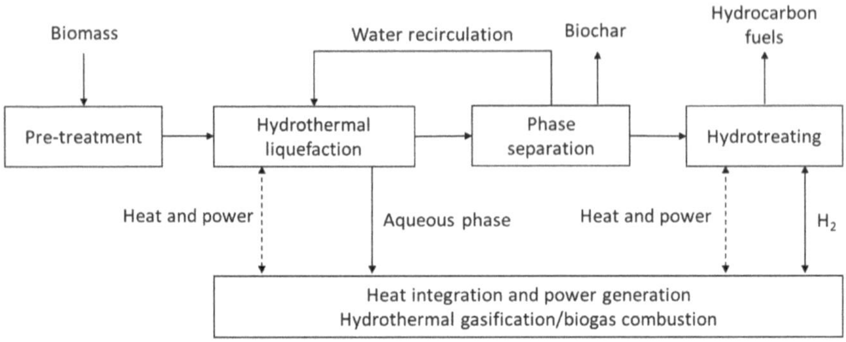

Figure 2.10 Process flow diagram of the integrated biomass hydrothermal liquefaction (HTL) process.

Fig. 2.10 shows a schematic diagram of the integrated biomass HTL process adapted with some modifications from Biller and Ross [60]. The integrated system mainly consists of an HTL reactor, hydrotreating unit, and heat/power generation. Depending on biomass feedstocks condition, pretreatment may be required, especially for the woody biomass to reduce the particle size, remove the contaminants, and alkaline treatment. Subsequently, the feedstock is processed via HTL at a temperature of around 350 °C and a pressure of 150 bar. Phase separation occurs spontaneously at these operating conditions resulting in the gaseous stream of CO_2, solid residue, bio-crude, and small traces of aqueous phase [149]. The aqueous phase/water can be recirculated to the HTL unit, which reduces the water requirement and enhances the bio-oil yield. The water generated in the HTL process can be treated anaerobically or via catalytic hydrothermal gasification to produce methane-rich or hydrogen-rich syngas.

In the phase separation unit, the solid phase material can be directly used as biochar or fertilizer. The produced bio-crude may not be directly used for fuels, hence requires further hydrotreatment for commercial utilization. However, the upgrading does not require complicated procedures as the bio-crude obtained through HTL is less in moisture and oxygen content. Hence, the fine hydrotreatment will enhance the quality of bio-crude. Finally, the amount of heat generated through the entire process can efficiently operate the hydrothermal gasifier that refines the process economy.

2.2.5 Thermochemical cycle

Table 2.5 Several groups of hydrolytic enzymes and breakdown products [63].

Substrates	Enzymes	Breakdown products
Proteins	Proteinase	Amino acids
Cellulose	Cellulase	Cellobiose and glucose
Hemicellulose	Hemicellulase	Sugars, such as glucose, xylose, mannose, and arabinose
Starch	Amylase	Glucose
Fats	Lipase	Fatty acids and glycerol
Pectin	Pectinase	Sugars, such as galactose, arabinose, and polygalactic uronic acid

2.3 Biochemical conversion

2.3.1 Anaerobic digestion

Anaerobic digestion is a series of complex biochemical processes by which microorganisms break down organic materials and convert them into methane and CO_2 mixture in the absence of oxygen. It is used as part of the waste management process to treat biodegradable waste and sewage sludge as well as biogas production, offering a clean and renewable form of energy. Animal manure and slurry, agricultural waste, and dedicated energy crops are commonly used as biomass feedstocks for biogas production. The biogas mainly consists of methane, CO_2, and traces of contaminant gases, which can be used directly as fuel in combined heat and power gas engines or upgraded to natural gas—quality biomethane.

The anaerobic digestion process involves hydrolysis, acidogenic fermentation, acetogenesis, and methanogenesis [61]. In the first step, complex organic substances that cannot be directly utilized by bacteria are decomposed into soluble monomers such as monosaccharides, amino acids, and other simple organic compounds with the effect of extracellular hydrolytic enzymes. This step is carried out by obligate anaerobes such as Bacteroides, Clostridia, and other facultative bacteria. The rate of decomposition during the hydrolysis step depends mainly on the nature of the substrate. Cellulose decomposition and hemicellulose generally take place more slowly than the proteins [62,63]. Table 2.5 shows examples of some different groups of extracellular hydrolytic enzymes and their functions.

In the second step, acidogenic fermentation bacteria convert the hydrolysis products into short-chain organic acids such as butyric acids, propanoic acids, acetic acids, alcohols, hydrogen, and CO_2. In general, during this phase, simple sugars, fatty acids, and amino acids are converted into acetate, CO_2, and hydrogen (70%) as well as into volatile fatty acids (VFAs) and alcohols (30%). Furthermore, products from acidogenesis, which cannot be directly converted to methane by methanogenic bacteria, are converted into methanogenic substrates during acetogenesis. The process involves oxidation reactions of the VFA and alcohol by microorganisms, which produce methanogenic substrates like acetate, hydrogen, and CO_2. The process requires close cooperation between the organisms that carry out oxidation and the methane-producing microorganisms active in the next stage. Finally, the production of methane and CO_2 is carried out by methanogenic bacteria. 70% of the formed methane originates from acetate, while the remaining 30% is produced from the conversion of hydrogen and CO_2 [64]. Methanogenesis is considered a critical step in the entire anaerobic digestion process, as it is the slowest biochemical reaction of the process. It is a complex process carried out by the synergistic action of various mesophilic bacterial species. The simplified diagram of four main steps during the anaerobic digestion process is shown in Fig. 2.11.

Anaerobic digestion depends on several different parameters for optimum performance, such as temperature, pH, C/N ratio, amount of volatile solids, retention time, and mixing [62,65]. It is considered a slow process since microorganisms need to adapt to new conditions (substrate, temperature, etc.). For example, biogas yield decreases at C/N ratio below 20 [66], initial TS over 30% [67], HRT below or above 15–30 days [68], and pH below or above 6.6–7.4 [69]. Anaerobic digestion can be conducted as a batch process or a continuous process. In a batch system, the feedstock is added to the reactor at the start of the process. This system has a simple

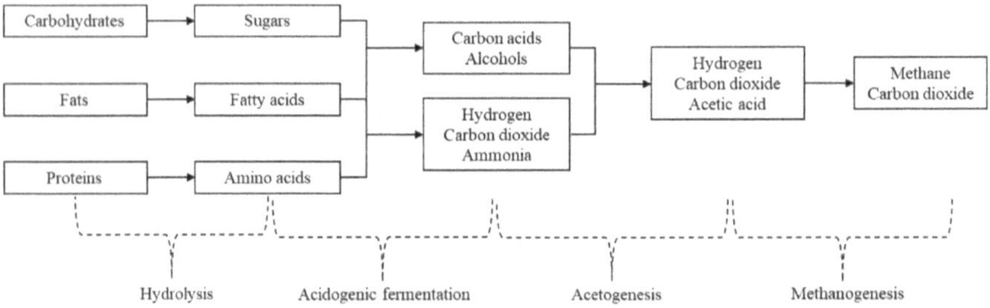

Figure 2.11 The simplified diagram of the main steps of the anaerobic digestion.

design and requires less equipment. Constant biogas production can be achieved by utilizing more than one batch reactor in a biogas plant. In a continuous system, biomass feedstocks are added continuously to the reactor. The biogas product is periodically removed, resulting in the constant production of biogas. In the continuous operation, single or multiple digesters in sequence may be used. This configuration can adopt up-flow anaerobic sludge blankets, expanded granular sludge beds, and internal circulation reactors.

The anaerobic digestion can be performed in a one-stage or multistage process. The one-stage anaerobic digestion is the simplest system that allows all steps (hydrolysis, acidogenesis, acetogenesis, and methanogenesis) to occur in a single reactor. The most complicated of this system is how to keep microorganisms in balance during the whole process. As reported, two groups of microorganisms differ widely in terms of physiology, nutritional needs, growth kinetics, and sensitivity toward environmental conditions [70]. A multistage process uses two or more reactors for digestion to separate the hydrolysis/acidogenesis and acetogenesis/methanogenesis phases. Acidogenic bacteria produce organic acids and more quickly grow and reproduce than methanogenic bacteria. Methanogenic bacteria require a stable pH and temperature to optimize their performance [71]. The multistage process is adopted to optimize every digestion step and achieve more stable operation, higher organic loading capacity, and a higher resistance toward toxicants and inhibiting substances [72–74]. Consequently, this technology requires higher investment and operation costs than a one-stage method. The configuration of the two-stage digestion system is shown in Fig. 2.12.

2.3.2 Fermentation

Fermentation is the core of the bioethanol production process, in which the sugars from organic wastes are converted to bioethanol. The organic material can be obtained from food waste, agricultural waste, sewage sludge, or other biomass waste. Before fermentation, some wastes require saccharification or hydrolysis to obtain a high concentration of glucose. Fermentation can then convert glucose to ethanol by microbes, such as *Saccharomyces cerevisiae*, *Escherichia coli*, *Zymomonas mobilis*, *Pachysolen tannophilus*, and *Candida shehatae* [75]. Although the fermentation has similarities with anaerobic digestion, fermentation products are alcohol or organic acids instead of methane. The overall yield of the process depends on the proportions of cellulose, hemicellulose, and lignin in the organic materials. As a

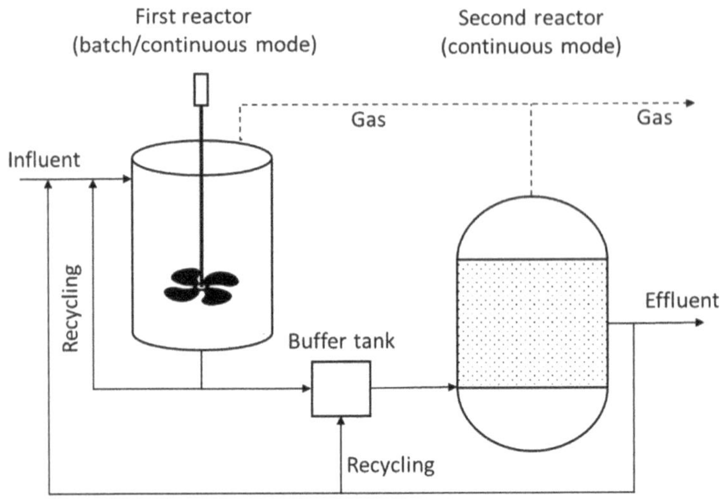

Figure 2.12 The configuration of the multistage digestion system using two reactors.

biological process, bioethanol fermentation is also dependent on many factors, including pH, oxygen, and temperature.

Sugar crops, starch crops, and lignocellulosic biomass have been used as raw materials on a commercial scale for bioethanol production to be used as an automotive fuel after further distillation. This large-scale bioethanol production basically involves several main steps, such as fermentation of sugars, distillation, dehydration, and denaturation (optional). The biodegradation of organic materials starts during the hydrolysis of complex compounds into simple compounds. The fermentation step is conducted by microbes that ferment sugars into ethanol, lactic acid, or other products. The ethanol is then separated from water, utilizing a fractional distillation process by exploiting the difference between the boiling point of water (100°C) and ethanol (78.3°C). Furthermore, the water can be separated using a condensation process, and ethanol distillate is recaptured at a concentration of 95% [76].

2.3.3 Photobiological H_2 production

Solar energy can be harvested to hydrogen through photosynthesis catalyzed by hydrogenases. This method is known as photobiological or biophotolysis, or photofermentation. The photobiological process converts organic compounds to H_2, involving various bacteria photosynthetic groups by a series of biochemical reactions [77]. The process can be performed in various algae and bacteria species and requires only light and CO_2 without additional substrates as nutrients. Hydrogen is split from the water via photosynthesis using sunlight and CO_2 as energy sources. Sunlight and CO_2 are the primary inputs to grow the organisms on the photobiological process supported by the hydrogenase enzyme. Fig. 2.13 shows the simplified schematic of photobiological hydrogen production in algae or bacteria and its utilization in the fuel cell system to produce electricity. This mechanism produces only water vapor and power at the point of use. H_2 can also be stored, making the use of hydrogen highly attractive as an energy carrier.

The photobiological process for hydrogen production can be classified into direct and indirect conversion. In direct photobiological conversion, hydrogen can directly be produced from water using organisms involved, such as microalgae. The hydrogen gas is produced using solar energy to split water molecules. However, this process also decreases ferredoxin, hydrogenase, or nitrogenase, in which these compounds are very sensitive to oxygen. The following reaction generally occurs during photosynthesis,

$$2H_2O + \text{solar energy} \rightarrow 2H_2 + O_2$$

In indirect photobiological, hydrogen production occurs in two phases. In the first phase, photosynthesis and carbohydrate storage take place. In the second phase, fermentation of stored carbohydrates leads to the production of hydrogen. The process of indirect photobiological conversion is shown in the following reactions:

$$12H_2O + 6CO_2 + \text{light} \rightarrow C_6H_{12}O_6 + 6O_2$$
$$C_6H_{12}O_6 + 12H_2O \rightarrow 12\,H_2 + 6CO_2$$

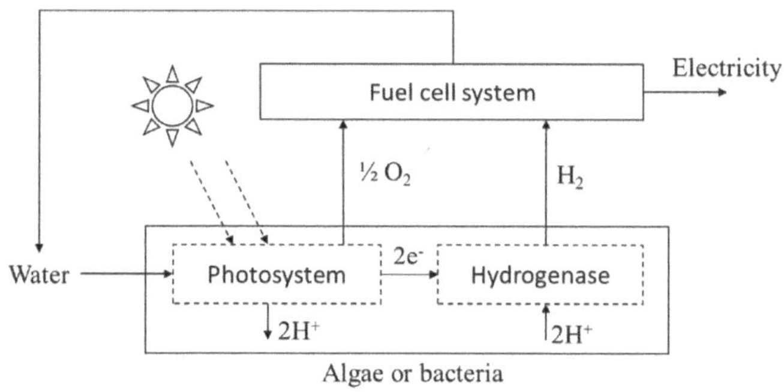

Figure 2.13 Schematic of the photobiological process for H_2 production and its utilization.

Depending on oxygen generation during photosynthesis, this process can be either anoxygenic or oxygenic. Anoxygenic photosynthesis occurs in some organisms by taking electrons from electron donor substrates like H_2S. The mechanism of anoxygenic photosynthesis or photofermentation through nitrogenase action is shown in Fig. 2.14, adapted from Ref. [109]. Due to the inability to get electrons from water, anoxygenic photosynthesis is not applicable on a large scale and not feasible commercially [78]. The oxygenic photosynthesis occurs in plants like microalgae and cyanobacteria by utilizing solar to split water into oxygen and H_2. The presence of generated oxygen during the process is a significant challenge for bio-hydrogen production [79]. The presence of oxygen can inhibit the enzyme function, transcription, and protein maturation [80].

Researchers are trying to optimize biohydrogen production and overcome the presence of oxygen molecules. Melis [81] has proven that the enzyme hydrogenase activity requires the absence of oxygen in the algae culture, such as *Chlamydomonas*. Paramesh and Chandrasekhar [82] reported the use of oxygen scavengers, namely sodium sulfite, sodium metabisulfite, and sodium dithionite, to enhance H_2 production from green algae. They found all of three scavengers could improve H_2 production. Sodium sulfite showed the best result due to its hyperactive oxygen consumption nature. Previously, $NaHSO_3$ has been tested to enhance hydrogen production in *Chlamydomonas reinhardtii* [83]. The result showed that a moderate amount of sodium bisulfite in the tested algae could remove O_2 efficiently. The

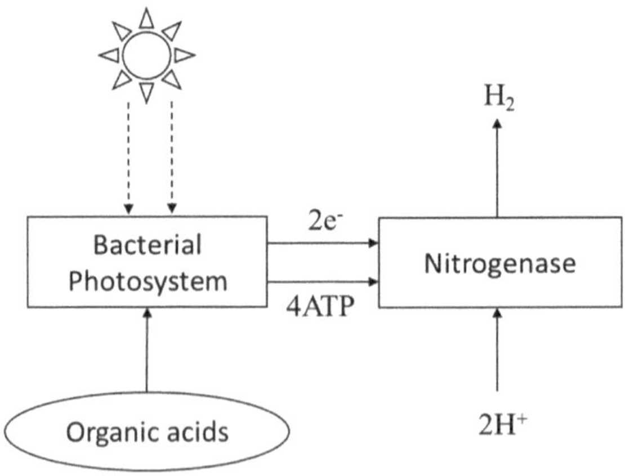

Figure 2.14 Schematic of anoxygenic photosynthesis or photofermentation through the action of nitrogenase.

utilization of oxygen blockers using the copper element to improve hydrogen production was also investigated in algae culture by Raymond et al. [84].

Another crucial factor in building a sustainable photobiological process for hydrogen production is effective photobioreactor and genetic engineering. The latter is considered as pioneering research to enhance H_2 yield during the process [79]. Genetic engineering has been applied in some photosynthetic bacteria, such as cyanobacteria and *Rhodobacter sphaeroides* [85–87]. Since most photobiological hydrogen conversion is investigated on a laboratory scale, the reactor's practical design is required on a larger scale. Some key parameters to design bioreactors are temperature control, agitation system, bioreactor performance with the same volume but different area–volume ratio, and capability to accommodate consortium organisms [88].

2.4 Correlated technologies

The products from the thermochemical or biochemical processes are usually combusted as heat sources in the conversion process (reuse), and/or to generate electricity. Intermediate products such as pyrolysis gases or syngas from gasification can be used for producing different types of fuels. However, the products should be applicable in some markets and also economically competitive.

2.4.1 Steam reforming

Steam reforming is the method to produce hydrogen and carbon monoxide gases by reacting to the hydrocarbons, especially methane, with steam. The hydrocarbons (syngas) are produced through gasification and pyrolysis. The basic reaction occurs as follows,

$$CH_4 + H_2O \rightleftharpoons CO + H_2$$

As could be observed from the reaction's heat, this reaction is endothermic; therefore, heat supply is needed to facilitate the reaction to occur. In general, the reaction is carried out inside the pressurized reactor with the assistant of a catalyst (such as nickel). In order to improve the conversion rate, the catalyst having a high surface area to volume ratio is strongly expected to minimize the limitation due to high temperature. However, the shape of the catalyst also should be able to facilitate a low-pressure drop.

Steam reforming is usually coupled with the water-gas shift reaction to enrich the produced hydrogen by converting the carbon monoxide to hydrogen. The water-gas shift reaction occurs as follow,

$$CO + H_2O \rightleftharpoons CO_2 + H_2$$

As the basic steam reforming process is endothermic, to cover the heat required during the reaction, partial oxidation of the fuel can be carried out. This method is generally known as autothermal reforming. This autothermal reforming can be conducted with CO_2 or steam as reactants. In the case of CO_2 is adopted as a reactant, the reaction occurs as follow,

$$2CH_4 + O_2 + CO_2 \rightleftharpoons 3CO + 3H_2 + H_2O$$

Furthermore, in case that steam is supplied as the reactant, the reaction proceeds as follows,

$$2CH_4 + O_2 + 2H_2O \rightleftharpoons 4CO + 10H_2$$

2.4.2 Water-gas shift reaction

Water-gas, also known as synthesis gas (syngas), contains carbon monoxide (CO) and hydrogen (H_2). Since the gasification produces large amounts of carbon monoxide and hydrogen (with a small number of other gases), the water-gas shift process is necessary to obtain more hydrogen by reacting the carbon monoxide and steam. Subsequently, the H_2-rich stream can be obtained after CO_2 removal by a simple alkaline wash. The gasification, coupled with a water-gas shift, is the most widely practiced process route for biomass to H_2 [89–91]. The water-gas shift process, a reversible exothermic reaction, occurs as follow,

$$CO_{(g)} + H_2O \leftrightarrow CO_{2(g)} + H_{2(g)}$$

Not only in gasification and steam reforming processes, but the water-gas shift is also encountered in the reactions of steam reforming process [92,93], partial oxidation [94,95], autothermal reforming [96,97], methanol and dimethyl ether (DME) syntheses [98,99], and the Fischer–Tropsch process [100]. Although the water-gas shift reaction is not classified as one of the primary thermochemical conversions, it cannot be omitted to analyze chemical reaction systems that involve syngas. Among reactions involving syngas, this reaction equilibrium is least sensitive to the temperature change [101].

The water-gas shift reaction can be performed with or without a catalyst. The classification of these processes is shown in Fig. 2.15. In the noncatalytic process, a particular environment is required, such as supercritical water and plasma systems. Conventionally, iron-chromium (Fe—Cr) and copper-zinc (Cu—Zn) catalysts have been used during the water-gas shift reaction at high (350—500°C) and low temperatures (200—250°C), respectively. Pal et al. [102] reported the development of catalyst technology to facilitate the reaction and increase the CO conversion. To date, the investigations are mostly focused on ceria (CeO_2) based catalysts due to its favorable properties. They concluded that the current research in this area is vast, with improvements and breakthroughs in the near future.

The water-gas shift process has some challenges to be overcome, such as catalyst mechanism, cost-efficient catalysts, and a comprehensive study of the special environments for catalytic or noncatalytic processes. Another critical part being addressed is carbon capture/storage and membrane systems to purify the hydrogen yield and enrich CO_2 [103]. Moreover, this shifting method can increase the entire conversion chain's efficiency from biomass or waste into hydrogen product.

2.4.3 Gas separation

Pressure swing adsorption (PSA) is employed to separate the impurities to produce high hydrogen stream quality. The raw hydrogen gas supplied to PSA can be obtained from various processes such as steam methane reforming, partial oxidation, cryogenic purification, gasification, and ammonia plants. It physically utilizes adsorbent material to bind the gas molecules, which depends on the gas component, type of adsorbent material, the gas component's partial pressure, and operating temperature [104]. The raw hydrogen is processed and purified at a pressure that closes

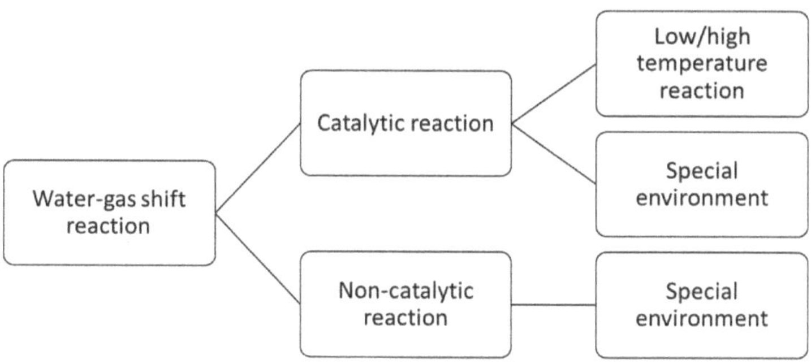

Figure 2.15 Classification of the water-gas shift reaction.

to feeding pressure, while adsorbed impurities are recovered in the lower pressure. The process can generate hydrogen up to 99.99% and typically has a hydrogen separation efficiency of 70%–85%.

In the PSA, the effect of separation is based on differences in binding forces to the adsorbent material. Since hydrogen has a low polarity, it is practically nonadsorbable material. PSA work by exploiting the adsorbable characteristic of molecules other than hydrogens such as nitrogen, carbon monoxide, CO_2, hydrocarbon derivatives, and water vapor. Therefore, these impurities can be removed at the lower pressure from a raw hydrogen stream with higher pressure, and high-purity hydrogen is recovered.

Several studies involving PSA units have been reported for hydrogen production from biomass. Fig. 2.16 shows an overview of hydrogen production routes from biomass, including the PSA's gas separation process. Vera et al. simulated an integrated system consisting of gasification, water-gas shift reactor, and PSA unit using the Aspen Plus modeler [105]. The results showed that after replacing the PSA unit with the palladium membrane, the process's hydrogen recovery increased from 38% to 49%. Unfortunately, the economic impact of the use of the membrane was not taken into consideration.

Another route of hydrogen production was also studied involving the gasification, steam reforming, and PSA unit. The study indicated this combination has a higher energy efficiency compared to the biomass-gasification-electricity-electrolysis route. However, it requires improvements in its environmental efficiency based on the life cycle assessment (LCA) [106]. Cohce et al. [107] performed energy and exergy analyses of a biomass-based hydrogen production system. The system consists of gasification, steam methane reforming, shift reaction, including PSA unit. They reported the exergy and energy efficiency of 19% and 22%, respectively. The study also indicated that exergy analysis is a useful tool for understanding and improving efficiency. In the PSA unit system, physical exergy losses due to temperature and pressure differences between the inlet and

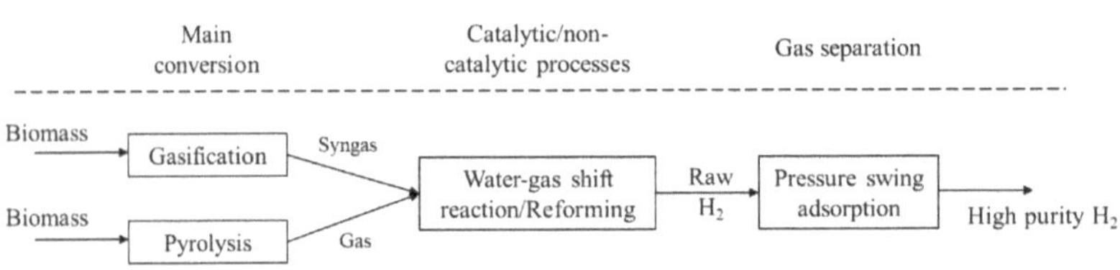

Figure 2.16 Schematic of the main processes involving gas separation.

outlet gases. Jurgen et al. [108] also reported the plant's mass and energy balances consisting of a dual fluidized, a water gas shift, and a PSA unit for hydrogen production. The validation showed that chemical and energy efficiency of about 60% and 55%, respectively, indicated this hydrogen production route's high potential.

2.4.4 Liquid hydrocarbons via Fischer Tropsch

Fischer Tropsch is a gas-to-liquid polymerization technique to convert carbon monoxide and hydrogen into liquid hydrocarbon fuels for petroleum product substitution. It can be coupled with the gasification process to obtain the syngas (CO and H_2) before being converted into liquid fuels. The biomass sources are converted to *syngas* through gasification, where a controlled flow of steam and oxygen is maintained through the source at high temperature and pressure. Fischer Tropsch is an exothermic process by utilizing a metal catalyst which causes polymerization, according to the following equation,

$$nCO + (2n + 1)H_2 \rightarrow CH_{2n+2} + nH_2O$$

Depending on some parameters including the catalyst, temperature (typically carried out in the range 200–350°C and 20–300 atm), and type of process employed, the Fischer Tropsch can produce hydrocarbons ranging from methane (an undesirable by-product) to higher molecular paraffin and olefins. The most common metal catalyst consists of iron, cobalt, or ruthenium. The first two (Fe and Co) are most frequently used. Nickel can also be employed but tends to favor methane formation.

Since the process is highly exothermic, it is important to design a reactor that considers the heat removal aspect, especially in a commercial operation. In order to control the reaction temperature, the heat is removed by steam generation. The energy recovery is considered necessary because 20%–25% of the syngas' heating value is released as heat. Three standard reactors can be used for the process and have been used commercially, such as fixed bed, fluidized bed, and slurry bed reactor. The fluidized bed reactors typically are operated in the temperature range 320–350°C [109]. This temperature range is 100°C higher than the typical operating temperature range used with the fixed bed or slurry bed reactors. These two configurations operate at a lower temperature of around 220–250°C.

References

[1] Bajpai P. Biomass conversion processes. Biomass to energy convers. Technol., Elsevier; 2020. p. 41–151. https://doi.org/10.1016/b978-0-12-818400-4.00005-0.
[2] Ptasinski KJ. Efficiency of biomass energy - an exergy approach to biofuels. Power Biorefineries 2017;89. https://doi.org/10.1002/cite.201770067.
[3] Hall CW, Mujumdar AS, Molnár K, Pakowski Z, Marinos-Kouris D, Maroulis ZB, et al. Handbook of industrial drying. Crc 2006;6:1279. https://doi.org/10.1080/07373938808916399.
[4] Pang S, Mujumdar AS. Drying of woody biomass for bioenergy: drying technologies and optimization for an integrated bioenergy plant. Dry Technol 2010;28:690–701. https://doi.org/10.1080/07373931003799236.
[5] Verma M, Loha C, Sinha AN, Chatterjee PK. Drying of biomass for utilising in co-firing with coal and its impact on environment – a review. Renew Sustain Energy Rev 2017;71:732–41. https://doi.org/10.1016/j.rser.2016.12.101.
[6] Li H, Chen Q, Zhang X, Finney KN, Sharifi VN, Swithenbank J. Evaluation of a biomass drying process using waste heat from process industries: a case study. Appl Therm Eng 2012;35:71–80. https://doi.org/10.1016/j.applthermaleng.2011.10.009.
[7] Darmawan A, Hardi F, Yoshikawa K, Aziz M, Tokimatsu K. Enhanced process integration of black liquor evaporation, gasification, and combined cycle. Appl Energy 2017;204:1035–42. https://doi.org/10.1016/j.apenergy.2017.05.058.
[8] Aziz M, Prawisudha P, Prabowo B, Budiman BA. Integration of energy-efficient empty fruit bunch drying with gasification/combined cycle systems. Appl Energy 2015;139:188–95. https://doi.org/10.1016/j.apenergy.2014.11.038.
[9] Zaini IN, Nurdiawati A, Aziz M. Cogeneration of power and H_2 by steam gasification and syngas chemical looping of macroalgae. Appl. Energy 2017. https://doi.org/10.1016/j.apenergy.2017.06.071.
[10] Holmberg H. Biofuel drying as a concept to improve the energy efficiency OF an industrial CHP plant. 2007.
[11] Yan W, Acharjee TC, Coronella CJ, Vásquez VR. Thermal pretreatment of lignocellulosic biomass. Environ Prog Sustain Energy 2009;28:435–40. https://doi.org/10.1002/ep.10385.
[12] Deng J, Wang GJ, Kuang JH, Zhang YL, Luo YH. Pretreatment of agricultural residues for co-gasification via torrefaction. J Anal Appl Pyrolysis 2009;86:331–7. https://doi.org/10.1016/j.jaap.2009.08.006.
[13] Chen WH, Kuo PC. A study on torrefaction of various biomass materials and its impact on lignocellulosic structure simulated by a thermogravimetry. Energy 2010;35:2580–6. https://doi.org/10.1016/j.energy.2010.02.054.
[14] Gebreegziabher T, Oyedun AO, Hui CW. Optimum biomass drying for combustion - a modeling approach. Energy 2013;53:67–73. https://doi.org/10.1016/j.energy.2013.03.004.
[15] Devolatilization - an overview | ScienceDirect Topics. n.d. Available from: https://www.sciencedirect.com/topics/engineering/devolatilization. (Accessed 9 July 2020).
[16] Williams A, Jones JM, Ma L, Pourkashanian M. Pollutants from the combustion of solid biomass fuels. Prog Energy Combust Sci 2012;38:113–37. https://doi.org/10.1016/j.pecs.2011.10.001.
[17] Sahu SG, Chakraborty N, Sarkar P. Coal-biomass co-combustion: an overview. Renew Sustain Energy Rev 2014;39:575–86. https://doi.org/10.1016/j.rser.2014.07.106.
[18] Sansaniwal SK, Pal K, Rosen MA, Tyagi SK. Recent advances in the development of biomass gasification technology: a comprehensive review. Renew Sustain Energy Rev 2017;72:363–84. https://doi.org/10.1016/j.rser.2017.01.038.

[19] Chopra S, Jain AK. A review of fixed bed gasification systems for biomass. 2007.
[20] Parametric Study of Fixed Bed Biomass Gasifier: A review - Inpressco. n.d. Available from: http://inpressco.com/parametric-study-of-fixed-bed-biomass-gasifier-a-review/. (Accessed 13 July 2020).
[21] Tańczuk M, Junga R, Werle S, Chabiński M, Ziółkowski L. Experimental analysis of the fixed bed gasification process of the mixtures of the chicken manure with biomass. Renew Energy 2019;136:1055–63. https://doi.org/10.1016/j.renene.2017.05.074.
[22] Lenis YA, Pérez JF, Melgar A. Fixed bed gasification of Jacaranda Copaia wood: effect of packing factor and oxygen enriched air. Ind Crops Prod 2016;84:166–75. https://doi.org/10.1016/j.indcrop.2016.01.053.
[23] Horttanainen M, Saastamoinen J, Sarkomaa P. Operational limits of ignition front propagation against airflow in packed beds of different wood fuels. Energy Fuels 2002;16:676–86. https://doi.org/10.1021/ef010209d.
[24] Lv P, Yuan Z, Ma L, Wu C, Chen Y, Zhu J. Hydrogen-rich gas production from biomass air and oxygen/steam gasification in a downdraft gasifier. Renew Energy 2007;32:2173–85. https://doi.org/10.1016/j.renene.2006.11.010.
[25] James RA, Yuan W, Boyette M. The effect of biomass physical properties on top-lit updraft gasification of woodchips. Energies 2016;9:283. https://doi.org/10.3390/en9040283.
[26] Antonopoulos I-S, Karagiannidis A, Elefsiniotis L, Perkoulidis G, Gkouletsos A. Development of an innovative 3-stage steady-bed gasifier for municipal solid waste and biomass. Fuel Process Technol 2011. https://doi.org/10.1016/j.fuproc.2011.08.016.
[27] Niu M, Huang Y, Jin B, Wang X. Oxygen gasification of municipal solid waste in a fixed-bed gasifier. Chin J Chem Eng 2014;22:1021–6. https://doi.org/10.1016/j.cjche.2014.06.026.
[28] Thanapal SS, Annamalai K, Sweeten JM, Gordillo G. Fixed bed gasification of dairy biomass with enriched air mixture. Appl Energy 2012;97:525–31. https://doi.org/10.1016/j.apenergy.2011.11.072.
[29] Basu P. Tar production and destruction. Biomass gasif. Des. Handb., Elsevier; 2010. p. 97–116. https://doi.org/10.1016/b978-0-12-374988-8.00004-0.
[30] Basu P. Biomass gasification and pyrolysis. Elsevier Inc.; 2010. https://doi.org/10.1016/C2009-0-20099-7.
[31] Susastriawan AAP, Saptoadi H, Purnomo. Small-scale downdraft gasifiers for biomass gasification: a review. Renew Sustain Energy Rev 2017;76:989–1003. https://doi.org/10.1016/j.rser.2017.03.112.
[32] Beohar H, Gupta B, Sethi VK, Pandey M, Uit MPB, Bhopal MP. Parametric study of fixed bed biomass gasifier: a review, vol. 2; 2012.
[33] Motta IL, Miranda NT, Maciel Filho R, Wolf Maciel MR. Biomass gasification in fluidized beds: a review of biomass moisture content and operating pressure effects. Renew Sustain Energy Rev 2018;94:998–1023. https://doi.org/10.1016/j.rser.2018.06.042.
[34] Anukam A, Mamphweli S, Reddy P, Meyer E, Okoh O. Pre-processing of sugarcane bagasse for gasification in a downdraft biomass gasifier system: a comprehensive review. Renew Sustain Energy Rev 2016;66:775–801. https://doi.org/10.1016/j.rser.2016.08.046.
[35] Bain RL, Broer K. Gasification. Thermochem. Process. Biomass convers. into fuels, chem. Power. John Wiley and Sons; 2011. p. 47–77. https://doi.org/10.1002/9781119990840.ch3.
[36] Ud Din Z, Zainal ZA. Biomass integrated gasification-SOFC systems: technology overview. Renew Sustain Energy Rev 2016;53:1356–76. https://doi.org/10.1016/j.rser.2015.09.013.

[37] Molino A, Chianese S, Musmarra D. Biomass gasification technology: the state of the art overview. J Energy Chem 2016;25:10–25. https://doi.org/10.1016/j.jechem.2015.11.005.
[38] Kaewluan S, Pipatmanomai S. Gasification of high moisture rubber woodchip with rubber waste in a bubbling fluidized bed. Fuel Process Technol 2011;92:671–7. https://doi.org/10.1016/j.fuproc.2010.11.026.
[39] Samiran NA, Jaafar MNM, Ng JH, Lam SS, Chong CT. Progress in biomass gasification technique - with focus on Malaysian palm biomass for syngas production. Renew Sustain Energy Rev 2016;62:1047–62. https://doi.org/10.1016/j.rser.2016.04.049.
[40] Puig-Arnavat M, Bruno JC, Coronas A. Modified thermodynamic equilibrium model for biomass gasification: a study of the influence of operating conditions. Energy and Fuels 2012;26:1385–94. https://doi.org/10.1021/ef2019462.
[41] Corella J, Toledo JM, Molina G. A review on dual fluidized-bed biomass gasifiers. Ind Eng Chem Res 2007;46:6831–9. https://doi.org/10.1021/ie0705507.
[42] Göransson K, Söderlind U, He J, Zhang W. Review of syngas production via biomass DFBGs. Renew Sustain Energy Rev 2011;15:482–92. https://doi.org/10.1016/j.rser.2010.09.032.
[43] Richardson Y, Drobek M, Julbe A, Blin J, Pinta F. Biomass gasification to produce syngas. Recent adv. Thermochem. Convers. Biomass. Elsevier Inc.; 2015. p. 213–50. https://doi.org/10.1016/B978-0-444-63289-0.00008-9.
[44] Basu P. Gasification theory. Biomass gasification, pyrolysis torrefaction pract. Des. Theory. Elsevier; 2018. p. 211–62. https://doi.org/10.1016/B978-0-12-812992-0.00007-8.
[45] Reed TB, Das A. Handbook of biomass downdraft gasifier engine systems. CO (United States): Golden; 1988. https://doi.org/10.2172/5206099.
[46] Hofbauer H, Materazzi M. Waste gasification processes for SNG production. In: Substit. Nat. Gas from waste tech. Assess. Ind. Appl. Biochem. Thermochem. Process. Elsevier; 2019. p. 105–60. https://doi.org/10.1016/B978-0-12-815554-7.00007-6.
[47] Akhtar J, Saidina Amin N. A review on operating parameters for optimum liquid oil yield in biomass pyrolysis. Renew Sustain Energy Rev 2012;16:5101–9. https://doi.org/10.1016/j.rser.2012.05.033.
[48] Lam MK, Loy ACM, Yusup S, Lee KT. Biohydrogen production from algae. Biohydrogen, Elsevier; 2019. p. 219–45. https://doi.org/10.1016/b978-0-444-64203-5.00009-5.
[49] Basu P. Biomass gasification, pyrolysis and torrefaction: practical design and theory. Elsevier; 2018. https://doi.org/10.1016/C2016-0-04056-1.
[50] Perkins G, Bhaskar T, Konarova M. Process development status of fast pyrolysis technologies for the manufacture of renewable transport fuels from biomass. Renew Sustain Energy Rev 2018;90:292–315. https://doi.org/10.1016/j.rser.2018.03.048.
[51] Yildiz G, Ronsse F, Duren R Van, Prins W. Challenges in the design and operation of processes for catalytic fast pyrolysis of woody biomass. Renew Sustain Energy Rev 2016;57:1596–610. https://doi.org/10.1016/j.rser.2015.12.202.
[52] Sharma A, Pareek V, Zhang D. Biomass pyrolysis - a review of modelling, process parameters and catalytic studies. Renew Sustain Energy Rev 2015;50:1081–96. https://doi.org/10.1016/j.rser.2015.04.193.
[53] Perkins G, Batalha N, Kumar A, Bhaskar T, Konarova M. Recent advances in liquefaction technologies for production of liquid hydrocarbon fuels from biomass and carbonaceous wastes. Renew Sustain Energy Rev 2019;115:109400. https://doi.org/10.1016/j.rser.2019.109400.
[54] Integrated Hydropyrolysis and Hydroconversion Process (IH2) for Production of Gasoline and Diesel Fuel From Wood, Algae, Cornstover, Lemna, and Bagasse

Feedstocks | AIChE. n.d. Available from: https://www.aiche.org/academy/videos/conference-presentations/integrated-hydropyrolysis-and-hydroconversion-process-ih2-production-gasoline-and-diesel-fuel-wood. (Accessed 17 July 2020).
[55] Marker TL, Felix LG, Linck MB, Roberts MJ. Integrated hydropyrolysis and hydroconversion (IH 2) for the direct production of gasoline and diesel fuels or blending components from biomass, part 1: proof of principle testing. Environ Prog Sustain Energy 2012;31:191–9. https://doi.org/10.1002/ep.10629.
[56] Melin K, Osmaa A, Strueven J, Eidam P, Appeln J, Anacker C, et al. From black liquor to second generation transportation fuels. 2019.
[57] Zhang Y, Chen WT. 5 - hydrothermal liquefaction of protein-containing feedstocks. Direct Thermochem. Liq. Energy Appl., Elsevier; 2018. p. 127–68. https://doi.org/10.1016/B978-0-08-101029-7.00004-7.
[58] Toor SS, Rosendahl L, Rudolf A. Hydrothermal liquefaction of biomass: a review of subcritical water technologies. Energy 2011;36:2328–42. https://doi.org/10.1016/j.energy.2011.03.013.
[59] Gollakota ARK, Kishore N, Gu S. A review on hydrothermal liquefaction of biomass. Renew Sustain Energy Rev 2018;81:1378–92. https://doi.org/10.1016/j.rser.2017.05.178.
[60] Biller P, Ross AB. Production of biofuels via hydrothermal conversion. Handb. Biofuels Prod. Process. Technol. 2nd ed. Elsevier Inc.; 2016. p. 509–47. https://doi.org/10.1016/B978-0-08-100455-5.00017-5.
[61] Li Y, Chen Y, Wu J. Enhancement of methane production in anaerobic digestion process: a review. Appl Energy 2019;240:120–37. https://doi.org/10.1016/j.apenergy.2019.01.243.
[62] Okolie KF. A review of biochemical process of anaerobic digestion. Adv Biosci Biotechnol 2015;6:205–12. https://doi.org/10.4236/abb.2015.63020.
[63] Schnürer A, Jarvis Å. Microbiological handbook for biogas plants Swedish waste management U2009:03 Swedish gas centre report 207. 2010.
[64] Al Seadi T, Rutz D, Prassl H, Kottner M, Finsterwalder T, Silke Volk RJ. Biogas handbook. 2008.
[65] Mao C, Feng Y, Wang X, Ren G. Review on research achievements of biogas from anaerobic digestion. Renew Sustain Energy Rev 2015;45:540–55. https://doi.org/10.1016/j.rser.2015.02.032.
[66] Ehimen EA, Connaughton S, Sun Z, Carrington GC. Energy recovery from lipid extracted, transesterified and glycerol codigested microalgae biomass. GCB Bioenergy 2009;1:371–81. https://doi.org/10.1111/j.1757-1707.2009.01029.x.
[67] Jewell WJ, Cummings RJ, Richards BK. Methane fermentation of energy crops: maximum conversion kinetics and in situ biogas purification. Biomass Bioenergy 1993;5:261–78. https://doi.org/10.1016/0961-9534(93)90076-G.
[68] Jankowska E, Sahu AK, Oleskowicz-Popiel P. Biogas from microalgae: review on microalgae's cultivation, harvesting and pretreatment for anaerobic digestion. Renew Sustain Energy Rev 2017;75:692–709. https://doi.org/10.1016/j.rser.2016.11.045.
[69] Brown D, Shi J, Li Y. Comparison of solid-state to liquid anaerobic digestion of lignocellulosic feedstocks for biogas production. Bioresour Technol 2012;124:379–86. https://doi.org/10.1016/j.biortech.2012.08.051.
[70] Demirel B, Yenigün O. Two-phase anaerobic digestion processes: a review. J Chem Technol Biotechnol 2002;77:743–55. https://doi.org/10.1002/jctb.630.
[71] Griffin ME, McMahon KD, Mackie RI, Raskin L. Methanogenic population dynamics during start-up of anaerobic digesters treating municipal solid waste and biosolids. Biotechnol Bioeng 1998;57:342–55. https://doi.org/10.1002/(SICI)1097-0290(19980205)57:3<342::AID-BIT11>3.0.CO;2-I.

[72] Van DP, Fujiwara T, Tho BL, Toan PPS, Minh GH. A review of anaerobic digestion systems for biodegradable waste: configurations, operating parameters, and current trends. Environ Eng Res 2020;25:1–17. https://doi.org/10.4491/eer.2018.334.
[73] Kumar RP, Bharathiraja B, Kataki R, Moholkar VS. Biomass Valor Bioenergy 2020. https://doi.org/10.1007/978-981-15-0410-5.
[74] Aslanzadeh S, Rajendran K, Taherzadeh MJ. A comparative study between single- and two-stage anaerobic digestion processes: effects of organic loading rate and hydraulic retention time. Int Biodeterior Biodegrad 2014;95:181–8. https://doi.org/10.1016/j.ibiod.2014.06.008.
[75] Rajendran K, Lin R, Wall DM, Murphy JD. Influential aspects in waste management practices. Sustain. Resour. Recover. Zero waste approaches. Elsevier; 2019. p. 65–78. https://doi.org/10.1016/b978-0-444-64200-4.00005-0.
[76] Limayem A, Ricke SC. Lignocellulosic biomass for bioethanol production: current perspectives, potential issues and future prospects. 2012. https://doi.org/10.1016/j.pecs.2012.03.002.
[77] Azwar MY, Hussain MA, Abdul-Wahab AK. Development of biohydrogen production by photobiological, fermentation and electrochemical processes: a review. Renew Sustain Energy Rev 2014;31:158–73. https://doi.org/10.1016/j.rser.2013.11.022.
[78] Allakhverdiev SI, Thavasi V, Kreslavski VD, Zharmukhamedov SK, Klimov VV, Ramakrishna S, et al. Photosynthetic hydrogen production. J Photochem Photobiol C Photochem Rev 2010;11:101–13. https://doi.org/10.1016/j.jphotochemrev.2010.07.002.
[79] Poudyal RS, Tiwari I, Koirala AR, Masukawa H, Inoue K, Tomo T, et al. Hydrogen production using photobiological methods. Compend. Hydrog. Energy. Elsevier; 2015. p. 289–317. https://doi.org/10.1016/b978-1-78242-361-4.00010-8.
[80] Cohen I, Knopf JA, Irihimovitch V, Shapira M. A proposed mechanism for the inhibitory effects of oxidative stress on Rubisco assembly and its subunit expression. Plant Physiol 2005;137:738–46. https://doi.org/10.1104/pp.104.056341.
[81] Melis A. Photosynthetic H_2 metabolism in *Chlamydomonas reinhardtii* (unicellular green algae). Planta 2007;226:1075–86. https://doi.org/10.1007/s00425-007-0609-9.
[82] Paramesh K, Chandrasekhar T. Improvement of photobiological hydrogen production in Chlorococcum minutum using various oxygen scavengers. Int J Hydrogen Energy 2020;45:7641–6. https://doi.org/10.1016/j.ijhydene.2019.05.216.
[83] Wei L, Yi J, Wang L, Huang T, Gao F, Wang Q, et al. Light intensity is important for hydrogen production in $NaHSO_3$-treated *Chlamydomonas reinhardtii*. Plant Cell Physiol 2017;58:451–7. https://doi.org/10.1093/pcp/pcw216.
[84] Surzycki R, Cournac L, Peltier G, Rochaix JD. Potential for hydrogen production with inducible chloroplast gene expression in *Chlamydomonas*. Proc Natl Acad Sci USA 2007;104:17548–53. https://doi.org/10.1073/pnas.0704205104.
[85] Bothe H, Schmitz O, Yates MG, Newton WE. Nitrogen fixation and hydrogen metabolism in cyanobacteria. Microbiol Mol Biol Rev 2010;74:529–51. https://doi.org/10.1128/mmbr.00033-10.
[86] Nguyen AV, Toepel J, Burgess S, Uhmeyer A, Blifernez O, Doebbe A, et al. Time-course global expression profiles of *Chlamydomonas reinhardtii* during photo-biological H_2 production. PloS One 2011;6:e29364. https://doi.org/10.1371/journal.pone.0029364.
[87] Kondo T, Arakawa M, Hirai T, Wakayama T, Hara M, Miyake J. Enhancement of hydrogen production by a photosynthetic bacterium mutant with reduced pigment. J Biosci Bioeng 2002;93:145–50. https://doi.org/10.1263/jbb.93.145.

[88] Dasgupta CN, Jose Gilbert J, Lindblad P, Heidorn T, Borgvang SA, Skjanes K, et al. Recent trends on the development of photobiological processes and photobioreactors for the improvement of hydrogen production. Int J Hydrogen Energy 2010;35:10218–38. https://doi.org/10.1016/j.ijhydene.2010.06.029.
[89] Chianese S, Loipersböck J, Malits M, Rauch R, Hofbauer H, Molino A, et al. Hydrogen from the high temperature water gas shift reaction with an industrial Fe/Cr catalyst using biomass gasification tar rich synthesis gas. Fuel Process Technol 2015;132:39–48. https://doi.org/10.1016/j.fuproc.2014.12.034.
[90] Kraussler M, Hofbauer H. Development and experimental validation of a water gas shift kinetic model for Fe-/Cr-based catalysts processing product gas from biomass steam gasification. Biomass Convers Biorefinery 2017;7:153–65. https://doi.org/10.1007/s13399-016-0215-9.
[91] Kirm I, Brandin J, Sanati M, Sanatia M. Shift catalysts in biomass generated synthesis gas. Topics Catal 2007. https://doi.org/10.1007/s11244-007-0236-5.
[92] Peppley BA, Amphlett JC, Kearns LM, Mann RF. Methanol-steam reforming on Cu/Zno/Al2O3. Part 1: the reaction network. Appl Catal A Gen 1999;179:21–9. https://doi.org/10.1016/S0926-860X(98)00298-1.
[93] Chein R, Chen YC, Chung JN. Numerical study of methanol-steam reforming and methanol-air catalytic combustion in annulus reactors for hydrogen production. Appl Energy 2013;102:1022–34. https://doi.org/10.1016/j.apenergy.2012.06.010.
[94] Agrell J, Boutonnet M, Fierro JLG. Production of hydrogen from methanol over binary Cu/ZnO catalysts: Part II. Catalytic activity and reaction pathways. Appl Catal A Gen 2003;253:213–23. https://doi.org/10.1016/S0926-860X(03)00521-0.
[95] Chen WH, Shen CT, Lin BJ, Liu SC. Hydrogen production from methanol partial oxidation over Pt/Al_2O_3 catalyst with low Pt content. Energy 2015;88:399–407. https://doi.org/10.1016/j.energy.2015.05.055.
[96] Deluga GA, Salge JR, Schmidt LD, Verykios XE. Renewable hydrogen from ethanol by autothermal reforming. Science (80-) 2004;303:993–7. https://doi.org/10.1126/science.1093045.
[97] Pasel J, Samsun RC, Tschauder A, Peters R, Stolten D. A novel reactor type for autothermal reforming of diesel fuel and kerosene. Appl Energy 2015;150:176–84. https://doi.org/10.1016/j.apenergy.2015.04.038.
[98] Graaf GH, Sijtsema PJJM, Stamhuis EJ, Joosten GEH. Chemical equilibria in methanol synthesis. Chem Eng Sci 1986;41:2883–90. https://doi.org/10.1016/0009-2509(86)80019-7.
[99] Huang MH, Lee HM, Liang KC, Tzeng CC, Chen WH. An experimental study on single-step dimethyl ether (DME) synthesis from hydrogen and carbon monoxide under various catalysts. Int J Hydrogen Energy 2015;40:13583–93. https://doi.org/10.1016/j.ijhydene.2015.07.168.
[100] Van Der Laan GP, Beenackers AACM. Intrinsic kinetics of the gas-solid Fischer-Tropsch and water gas shift reactions over a precipitated iron catalyst. Appl Catal A Gen 2000;193:39–53. https://doi.org/10.1016/S0926-860X(99)00412-3.
[101] Speight JG. Upgrading by gasification. Heavy oil recover. Upgrad., Elsevier; 2019. p. 559–614. https://doi.org/10.1016/b978-0-12-813025-4.00013-1.
[102] Pal DB, Chand R, Upadhyay SN, Mishra PK. Performance of water gas shift reaction catalysts: a review. Renew Sustain Energy Rev 2018;93:549–65. https://doi.org/10.1016/j.rser.2018.05.003.
[103] Chen WH, Chen CY. Water gas shift reaction for hydrogen production and carbon dioxide capture: a review. Appl Energy 2020;258:114078. https://doi.org/10.1016/j.apenergy.2019.114078.
[104] Speight JG. Hydrogen production. Heavy oil recover. Upgrad., Elsevier; 2019. p. 657–97. https://doi.org/10.1016/B978-0-12-813025-4.00015-5.

[105] Marcantonio V, De Falco M, Capocelli M, Bocci E, Colantoni A, Villarini M. Process analysis of hydrogen production from biomass gasification in fluidized bed reactor with different separation systems. Int J Hydrogen Energy 2019;44:10350–60. https://doi.org/10.1016/j.ijhydene.2019.02.121.
[106] Koroneos C, Dompros A, Roumbas G. Hydrogen production via biomass gasification-A life cycle assessment approach. Chem Eng Process 2008;47:1261–8. https://doi.org/10.1016/j.cep.2007.04.003.
[107] Cohce MK, Dincer I, Rosen MA. Energy and exergy analyses of a biomass-based hydrogen production system. Bioresour Technol 2011;102:8466–74. https://doi.org/10.1016/j.biortech.2011.06.020.
[108] Loipersböck J, Luisser M, Müller S, Hofbauer H, Rauch R. Experimental demonstration and validation of hydrogen production based on gasification of lignocellulosic feedstock. Chem Eng 2018;2:61. https://doi.org/10.3390/chemengineering2040061.
[109] Steynberg AP, Dry ME, Davis BH, Breman BB. Fischer-Tropsch reactors. Stud Surf Sci Catal 2004;152:64–195. https://doi.org/10.1016/s0167-2991(04)80459-2.

CHAPTER 3

Application of exergy analysis and enhanced process integration

Arif Darmawan[1] and Muhammad Aziz[2]
[1]Agency for the Assessment and Application of Technology (BPPT), Puspiptek Serpong, Tangerang Selatan, Indonesia; [2]Institute of Industrial Science, The University of Tokyo, Meguro-ku, Tokyo, Japan

This chapter is written with the objectives of deepening our undestanding on basic thermodynamics and, then, extending its application in exergy analysis and enhanced process integration. The essential background of engineering thermodynamics is explained first before discussing the exergy concept, pinch analysis, and enhanced process integration. Understanding thermodynamics concepts helps us in applying them in a targeted system to improve its efficiency. The enhanced process integration that we want to achieve is based on straightforward thermodynamics and; therefore, they can be used practically. There are four thermodynamics laws, which express empirical facts and define physical quantities, such as temperature, energy, heat, work, and entropy, which characterize the thermodynamic processes and systems in thermal equilibrium. In this book, the first and second laws of thermodynamics will mostly be referred to. The third law is negligible as it refers to the entropy of cryogenic systems near the absolute zero. The four laws of thermodynamics are [1,2] as follows:

- The **Zeroth Law of Thermodynamics** states that if two systems are in thermal equilibrium with a third system, they are also in thermal equilibrium. This law defines the temperature concept and makes thermometers possible.
- The **First Law of Thermodynamics** states the change in internal energy of a closed system equals the net heat transfer into the system, plus the network is done on the system. It means that energy is conserved; the only way the internal energy can be changed is through the transfer of energy by work or by heat. Perpetual motion machines of the first kind that produce work with no energy input are impossible.
- The **Second Law of Thermodynamics:** In a natural thermodynamic process, the sum of the interacting thermodynamic systems' entropies increases. Equivalently, perpetual motion machines of the second kind

Innovative Energy Conversion from Biomass Waste
ISBN 978-0-323-85477-1
https://doi.org/10.1016/B978-0-323-85477-1.00001-4

(machines that spontaneously convert thermal energy into mechanical work) are impossible.

This concept is required for system analysis since the principles of conservation do not suffice.

- The **Third Law of Thermodynamics** states that a system's entropy approaches a constant value as the temperature approaches zero. Except for noncrystalline solids (glasses), a system's entropy at absolute zero is typically close to zero. It is equal to the natural logarithm of the product of the quantum ground states.

There are so many examples of thermodynamic phenomena around us. In daily life, we are experiencing that the heat can be converted into work, and vice versa. Another example, by combusting solid biomass; the heat can also be obtained for various activities, such as cooking, material drying, and generating electricity. Heat can also be transferred between different objects. The laws of thermodynamics dictate these energy behaviors. Even our human bodies obey the laws of thermodynamics.

3.1 The first law of thermodynamics mass and energy rate balances for a steady flow process

The principle of conservation of energy is explained based on thermodynamics' first law. Energy is a fundamental thermodynamics concept and one of the most significant engineering analysis aspects [3]. The mass and energy balances discussed in this section are limited to simple systems. The first law of thermodynamics defines that the change in internal energy is equal to the difference of the quantity of heat (Q) supplied into a system and the work (W) done by the system. The first law is often formulated as:

$$E_1 - E_2 = Q - W \tag{3.1}$$

$$\Delta KE + \Delta PE + \Delta U = Q - W \tag{3.2}$$

The kinetic and potential energy are negligible assuming that there is no change in the kinetic or potential energy of the system. Then the energy balance becomes

$$\Delta U = Q - W \tag{3.3}$$

Fig. 3.1 shows an example of a thermodynamic system involving heat and work in a piston-cylinder system. When the gas inside the cylinder is heated, it will expand and push the piston up, thereby working on it.

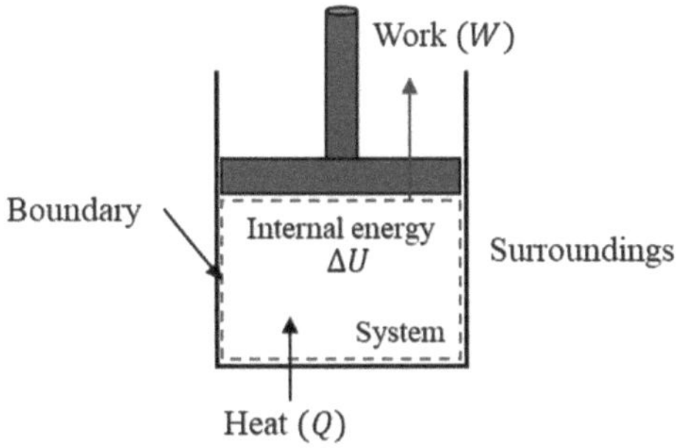

Figure 3.1 Example of a thermodynamic system of the gas confined by a piston in a cylinder.

However, the mode of heat transfer, whether by conduction, convection, or radiation, is not considered in this case. An increase in the system's internal energy results from positive heat transferred from the system's surroundings.

Work *(W)* due to compression or expansion can be seen in Fig. 3.1. The system is assumed to be a perfect closed system with no mass flow in or out, no friction losses, and quasi-equilibrium processes in which equilibrium conditions are managed. It is evaluated by integrating the force F multiplied by the incremental distance moved *dx* between an initial state (1) to a final state (2), shown as follows:

$$W_{1-2} = \int_{1}^{2} F dx \tag{3.4}$$

Systems can be either closed, opened, or isolated depending on the interaction with the surrounding. In an open system, the system can exchange both energy and matter with its surrounding. Fig. 3.2 illustrates a simple case of a one-inlet, one-outlet open system in a box container. A

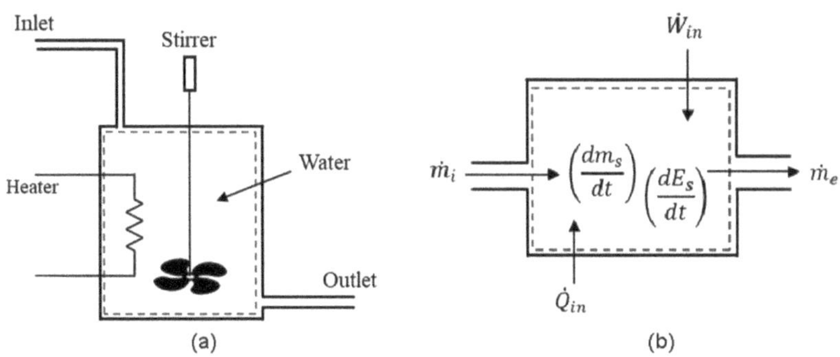

Figure 3.2 (A) Example of an open system in the container and (B) its block diagram.

heater and stirrer are placed in the system to heat and stir the water flowing through the container, respectively. The system boundary is indicated by a dashed line. However, the boundary of the system should exclude the stirrer and the heater, considering they are part of the surroundings. The container's mass and energy flow can be represented by a block diagram, as shown in Fig. 3.2B. The system consists of only the water in the container, marked by the system boundary. Water flows through the system and the terms $\dot{m}_i$ and $\dot{m}_e$ denote the rates at which mass enters and leaves the system, respectively, expressed in kg/s (SI units). If the increase in mass within the system is written by $\frac{dm_s}{dt}$, the law of conservation of mass becomes:

$$\frac{dm_s}{dt} = \dot{m}_i - \dot{m}_e \tag{3.5}$$

The mass balance, also called as a material balance, is the simplest application of conservation as the total mass cannot be created or destroyed. If chemical reactions occurred and the produced mass increases in a pollutant, they must also have caused a corresponding decrease in some other compounds' mass. The mass balance concept is widely applied to a steady and unsteady system, such as tanks, various reactors, turbines, pumps, and compressors. This principle is very beneficial and used in engineering and environmental analyses.

In a closed system, heat and work interactions between the system and its surroundings result in internal energy change. In an open system, the system's flow of matter is also considered as an additional mechanism affecting the internal energy change. Matter entering a system brings in some energy with it. Likewise, when matter leaves the system, it takes out some energy with it. Therefore, the energy balance for the open system shown in Fig. 3.2B includes heat, work, and energy of the mass streams entering and exiting the system as to the open system:

$$\frac{dE_s}{dt} = \dot{Q}_{in} + \dot{W}_{in} + \dot{E}_i - \dot{E}_e \tag{3.6}$$

where $\frac{dE_s}{dt}$ is the rate of increase in energy within the system. $\dot{E}_i$ and $\dot{E}_e$ are the rate at which energy is brought in and removed by the mass, respectively. The term $\dot{Q}_{in}$ represents the rate of energy transfer to the system from the heater, and the term $\dot{W}_{in}$ is the rate at which work enters the system from the stirrer. Integrating Eq. (3.6) with respect to time over the time interval Δt,

$$\Delta E_s = Q_{in} + W_{in} + \int_{t0}^{tf} \dot{E}_i dt - \int_{t0}^{tf} \dot{E}_e dt \tag{3.7}$$

where ΔE_s represents the change in the total energy of the system during Δt, and Q_{in} and W_{in} are amounts of net heat and net work entering the system during that period, respectively. In the steady-state open system, the equation can be written:

Steady

$$\frac{dE_s}{dt} = \dot{Q}_{in} + \dot{W}_{in} + \dot{m}_i\left(h_i + \frac{\vec{V}_i^2}{2} + gz_i\right) - \dot{m}_e\left(h_e + \frac{\vec{V}_e^2}{2} + gz_e\right) \tag{3.8}$$

where $\frac{\vec{V}^2}{2}$, gz, and h are the kinetic energy per unit mass (kJ/kg), the potential energy per unit mass (kJ/kg), and specific enthalpy, respectively. Using the assumption that kinetic and potential energy changes are negligible, we obtain:

$$\dot{Q}_{in} + \dot{W}_{in} = \dot{m}_i h_i - \dot{m}_e h_e \tag{3.9}$$

3.2 The second law of thermodynamics and entropy

The first law of thermodynamics affirms that energy cannot be created or destroyed. It means the energy quantity in the universe remains constant, but energy can be transferred from one object to another and transformed to and from other forms. The second law of thermodynamics is about the quality of energy as the first law's conservation principles do not always suffice. However, this second law is slightly confusing for most students and engineers as many complicated terms, such as enthalpy, entropy, Carnot's theorem, and Gibbs free energy. Most importantly, the majority of us do not understand what the applications of this law are.

The discussion about this law will be started with a simple case in our daily life. As shown in Fig. 3.3A, heat naturally flows from the hot (the coffee) to the cold (surrounding at 25°C) sides. It is spontaneous and occurs on its own. Heat does not naturally flow from the cold to the hot side. It can happen, at which heat does flow in the opposite direction from cold to hot, if we pump the energy into the system like in a refrigerator that takes

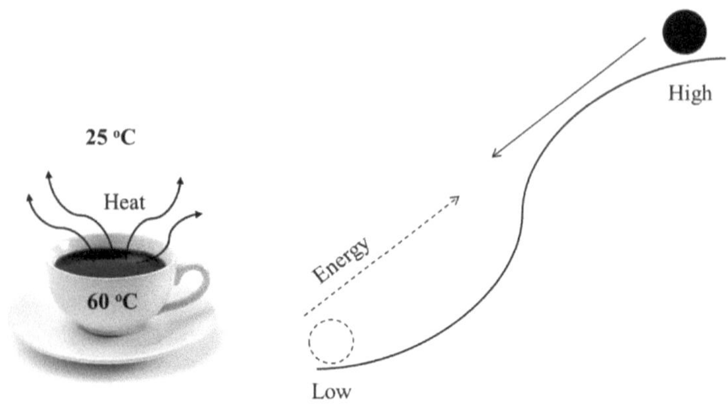

Figure 3.3 (A) The heat transfer from the coffee cup to the surrounding. (B) The ball on the top of the hill naturally flows in a downward direction.

out heat energy from inside the fridge and pumps it outside. In this way, the refrigerator can cool the stuff, but this phenomenon requires energy to do that and does not happen spontaneously. Another illustration is, we have a ball on the top of the hill (Fig. 3.3B). The ball will naturally flow in a downward direction in the same way the heat flows from the hot to the cold sides. We need to put energy into it to get the ball over the hill. The heat that naturally flows from hot to cold is a spontaneous process. For a spontaneous process, the entropy increases or is greater than zero $(S > 0)$. It is another statement of the second law, where for natural processes, the change in entropy will be greater than zero.

Let us imagine we have several core processes of industrial transformation, as seen in Fig. 3.4, such as heat transfer, flow, mass transfer, and

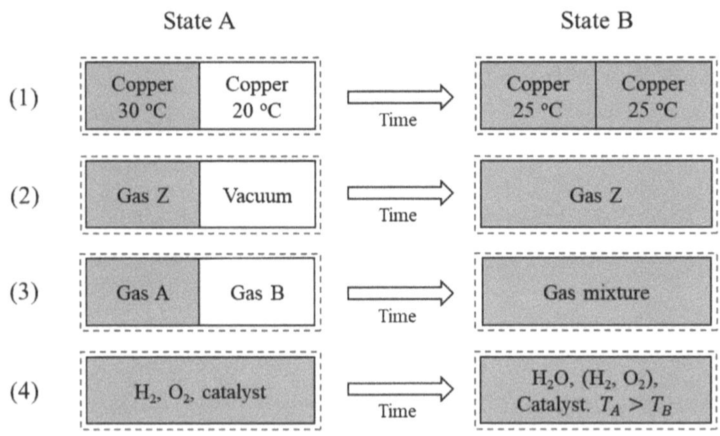

Figure 3.4 Entropy production in natural (irreversible) processes.

chemical reaction, as earlier discussed by Denbigh [4]. These processes are conducted in isolated systems, where there is no exchange of either energy or matter (mass) between the systems to the surroundings; therefore, both states A and B have the same mass and (internal) energy. The left side and the right side show a particular state A and state B of the system, respectively. State A or the initial state is defined by its temperature, volume, and composition. System (1) comprises two equal blocks of copper that connected each other. At the initial state, one block is at 25°C, and the other is at 30°C. As time goes by, heat conduction occurs and reaches the final state (state B), with both copper blocks have the same temperature of 25°C. System (2) has a container in which the half container has the initial state gas Z at temperature θ, and the other is empty. A wall separates this condition. After removing the wall, the gas Z expands and occupies the whole of the container at the same temperature θ. The system (c) also has a container that is equally separated by a wall, and each half is filled with gases A and B. After removing the wall, a uniform gas mixture occupies the throughout container space. Finally, the container in system (4) has oxygen, hydrogen, and a catalyst, at the initial state (A). The final state (B) contains water as a reaction product, unreacted oxygen and hydrogen, and the catalyst.

These processes in Fig. 3.4 naturally occur in a spontaneous process, as in Fig. 3.3. In each case, experience shows that it is impossible to carry out the process $B \rightarrow$ A spontaneously, although it is entirely consistent with the first law. The mass and internal energy remain the same for states A and B but differ in their quality. The distinction between the two states of the systems can be described using the concept entropy. The entropy of state B is always higher than the entropy in the respective state A: $S_A < S_B$ as the latter is characterized by a higher degree of molecular randomness or disorder [5]. The entropy difference can be estimated using standard thermodynamic relationships and properties of both states, such as volume, temperature, pressure, and concentration. In an open system, the changes in the amount of entropy contained within the system during some time interval (dS) can be expressed as a sum of two terms

$$dS = d_eS + d_iS \tag{3.10}$$

where d_eS is the net amount of entropy transferred across the system boundary and d_iS is the amount of entropy generated in the system due to irreversible processes, such as heat transfer, flow, mass transfer, and chemical reactions. The second law's entropy statement states, *It is impossible for any system to operate in a way that entropy is destroyed.* The entropy production term of Eq. (3.10) is always positive or zero and never negative. Thus, entropy production is an indicator of whether a process is possible or impossible.

The second law also has many important uses, such as [3]:

- To predict the direction of processes.
- To establish conditions for equilibrium.
- To determine the best theoretical performance of cycles, engines, and other devices.
- To evaluate quantitatively the factors that preclude the attainment of the best theoretical performance level.

Furthermore, the entropy analysis can provide a common platform to compare various biomass sources and the conversion technologies, e.g., pyrolysis, gasification, combustion [6]—the design results obtained from entropy and exergy analyses should be identical. In other words, the minimization of the total entropy generation is similar to the minimization of exergy destruction. Maximum exergy efficiency achieved in an investigated system would also be identical to the minimum entropy generation. The exergy concept and analysis will be discussed in the next subchapter.

Since the second law is based on the impossibility to 100% convert heat into work in an engine, its application in other devices should be addressed carefully [7]. Not all processes involve the presence of both heat and work like a refrigerator, a heat pump, or a heat engine, especially in the biomass utilization process. In the biomass conversion system, the calculation analysis solely on either entropy or exergy is insufficient. Therefore, one of this book's motivations is to give another perspective on improving the biomass utilization process. The exergy and entropy analyses could further be adapted to design industrial-scale plants or even individual devices to maximize energy/exergy efficiency and minimize their environmental impact. The efficient design plant means it has the best combination of feedstocks, processes, and products. Finally, in his notes in *Réflexions sur la puissance motrice du feu,* Carnot wrote: *We should not expect ever to utilize in practice all the motive power of combustibles. The attempts made to attain this result would be far more hurtful than useful if they caused other important considerations to be neglected. The economy of the combustible is only one of the conditions to be fulfilled in heat-engines. In many cases, it is only secondary. It should often give precedence to safety, to strength, to the durability of the engine, to the small space which it must occupy, to the small cost of installation, etc.*

3.3 Exergy concept

3.3.1 Exergy and energy

Exergy is the maximum theoretical work obtainable from an overall system consisting of a system and the environment as the system comes into equilibrium with the environment [8].

In the past, efficiency was measured and evaluated using mass and energy balance. There was no evaluation of material and energy quality involved in the system, and no environment reference used. In the 19th century, starting with Gibbs, L. Carnot and R. Clausius extended the concept of efficiency based on entropy and the second law of thermodynamics. In this approach, the quality of work and heat was considered to evaluate the system. In the 20th century, the concept of exergy was introduced based on the first and second thermodynamics law relating to energy and entropy, respectively. It was aimed to measure the quality of energy forms based on the environments as the reference system. Exergy can be defined the property that quantifies the potential for use. Its concept is used to evaluate the quality of energy since, in the real process, exergy is not conserved but consumed. Szargut used the following statement to explain the exergy term [9]: *Exergy is the amount of work obtainable when some matter is brought to a state of thermodynamic equilibrium with the common components of its surrounding nature by means of reversible processes, involving interaction only with the above mentioned components of nature.*

The reference environment is one of the most vital elements of exergy analysis. For example, we have a gas bottle full of air at 10 bar and 25°C. Using a nozzle, the pressure is converted into velocity, and the gas flow is accelerated to drive the turbine as a mechanical device. The work produced by the turbine can be used for generating electrical power. The next question is, what if the environmental pressure is 10 bar, as shown in Fig. 3.5B? It means the pressure inside the bottle is the same as the pressure outside the bottle. With this condition, there is not going to be any flow because the pressure is the same. Thus, we cannot generate electricity.

In the same reference environment, several systems can have different energy qualities. The different energy quality can refer to systems with the same energy form but differing in properties, such as temperature, pressure, and composition. See Fig. 3.6 with three systems with the same energy quantity but varying in their quality to illustrate this statement. These systems change over time. This example is taken from Moran and Shapiro [3]. Initially, an isolated system consists of a container of fuel surrounded by

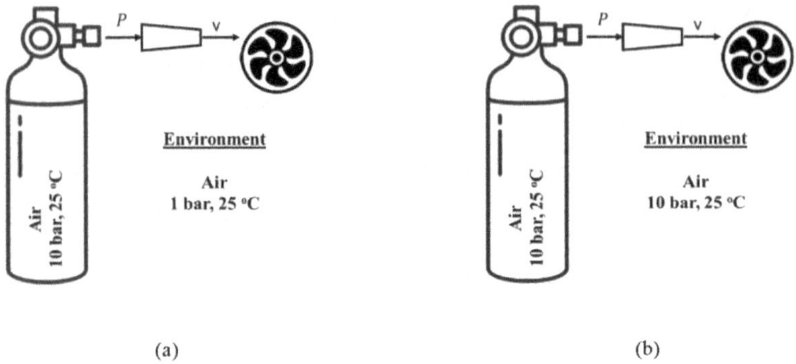

Figure 3.5 Two systems with different reference environments.

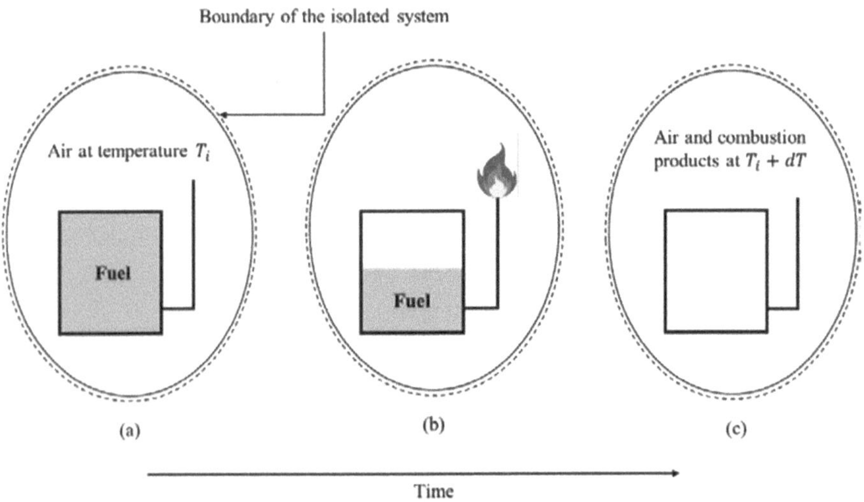

Figure 3.6 Three systems with the same quantity but the different quality of energy.

air at an initial temperature T_i (see Fig. 3.6A). After fuel burning as shown in the system (b) so in the final stage, there is a higher temperature of combustion products and air at a temperature $T_i + dt$, as seen in the system (c). The total energy associated with the system is constant because of no energy transfers across the boundary of an isolated system. The initial fuel-air combination at the system (a) is intrinsically more useful than the final warm mixture. It can be utilized for more purposes than a hot gas in the system (c), for instance, to generate electricity or produce superheated steam. In contrast, the uses of a mixture at the final system (c) are far more limited. We can conclude both the potential for use and economic value decrease from the system (a) to system (c) over time. More precisely, the initial potential is primarily destroyed because of the irreversible nature of the process.

Table 3.1 presents the comparison between the key properties of energy and exergy. Although energy is a correct scientific concept, it is sometimes confusing in a real application. Unlike energy, exergy depends on the properties of energy or matter flow and the environmental parameters where all real applications or industrial activities occur. It means the thermodynamic value of resources is estimated with reference to the environment. From this understanding, a difference with the environment plays an important role in generating exergy [10]. A simple illustration can be created for heat as a form of energy since it can be broken down into useful energy (its exergy) and one part that cannot produce work, which is referred to as anergy,

$$\text{Energy} = \text{exergy} + \text{exergy} \tag{3.11}$$

However, this simple equation can be applied to heat, and this kind of exergy will later be referred to as temperature-based exergy. The amount of work produced increases with the higher temperature, using the ambient temperature T_0 as the reference. If the pressure and chemical composition are considered, the simple relation in Eq. (3.11) can no longer be used [11]. Finally, the principal similarity between both concepts, energy and exergy, is that they are expressed in the same units, namely, joules (energy and exergy) or watts (energy and exergy rate). Exergy is measured in the same units as energy, although being a different concept because it is defined as "maximum work" (the ordered energy form).

Table 3.1 The comparison between energy and exergy.

Energy	Exergy
The first law of thermodynamics	The first and second law of thermodynamics
It cannot be destroyed	Consumed in a real process
Measure of quantity	Measure of quality
Defined only by properties of energy or matter flow	Defined by properties of energy or matter as well as by the environmental conditions
Expressed in energy units	Expressed in energy units

3.3.2 Classification of exergy

The exergy total can be divided into two primary forms, mechanical and thermal exergy (see Fig. 3.7). The mechanical exergy consists of kinetic and potential exergy. Since kinetic and potential exergy are evaluated relative to the environment, its exergies are zero. The thermal exergy is the sum of thermomechanical or physical exergy and chemical exergy. Thermomechanical exergy can be decomposed into a pressure-based part and a temperature-based part. Physical exergy (ex_{ph}) of a flow of matter is equal to the obtainable maximum work (W_{max}) when its state is changed from the initial state (at pressure P and temperature T) to the environmental state. Temperature- and pressure-based exergy can be manipulated individually by changing temperature and pressure, respectively. However, according to Truls [11], thermomechanical exergy decomposition into temperature-based or pressure-based part has no real fundamental significance and lack of a deeper meaning.

Although it is impossible to create a system with 100% efficiency in real engineering applications, Fig. 3.8A illustrates a model comprising an ideal system in which a matter flow is brought through reversible physical processes from the initial to the environmental state (P_0, T_0). Chemical exergy is equal to the maximum obtainable work from a stream when its

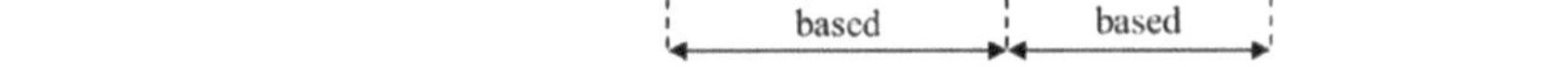

Figure 3.7 Exergy component breakdowns [12].

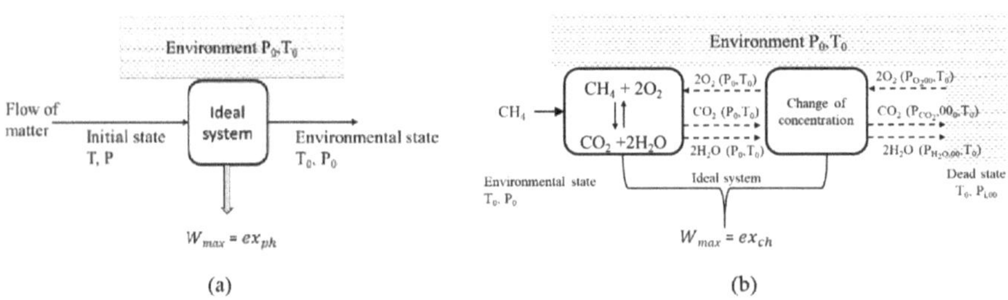

Figure 3.8 Schematic model to determine the flow of (A) physical and (B) chemical exergy.

state is changed from the environmental state (defined by the environmental pressure P_0 and temperature T_0) to the dead state that is characterized by P_0, T_0, and the reference chemical species [13].

Fig. 3.8B shows the schematic model of the CH_4 stream as an example to evaluate the chemical exergy flow. The maximum work (W_{max}) of the chemical exergy is equal to the standard change of free energy of the chemical reaction. The chemical reaction is a combustion process between methane and oxygen in the air. When combustion occurs, heat, carbon dioxide (CO_2), and water (H_2O) are produced.

$$CH_4 + 2O_2 = CO_2 + 2H_2O \tag{3.12}$$

For common chemical compounds, exergy value can be found in the book by Szargut et al. [14] to determine the chemical exergy of gaseous and liquid organic fuels using the standard chemical exergy tables. However, it is impossible to use it in most solid and liquid organic fuels consisting of complex solutions and mixtures of many (usually unknown) compounds. In order to overcome this limitation, Styrylska and Szargut [15] proposed a calculation method based on an analogy with the chemical exergy of purely organic substances. After calculating the chemical exergy of several organic substances, the approximate formula has been derived expressing the ratio β of their chemical exergy to the lower heating value as a function of the atomic ratio of the elements C, H, O, N, S.

$$\beta = \frac{ex_{ch,\ i}}{LHV_i} \tag{3.13}$$

where $ex_{ch,i}$ and LHV_i denote the chemical exergy and the lower heating value, respectively, of organic compound i. The most important formulae for solid:

$$\begin{aligned} \beta &= 1.0347 + 0.0140\frac{H}{C} + 0.0968\frac{O}{C} + 0.0493\frac{N}{C}; \\ \frac{O}{C} &< 0.5 \end{aligned} \tag{3.14}$$

$$\beta = \frac{1.044 + 0.0160\left(\frac{H}{C}\right) - 0.3493\left(\frac{O}{C}\right)\left[1 + 0.0531\left(\frac{H}{C}\right)\right] + 0.0493\left(\frac{N}{C}\right)}{1 - 0.4124\left(\frac{O}{C}\right)}$$

$$\frac{O}{C} < 2 \tag{3.15}$$

For liquid C, H, O, S compounds,

$$\beta = 1.047 + 0.0154\frac{H}{C} + 0.0562\frac{O}{C} + 0.5904\frac{S}{C}\left(1 - 0.175\frac{H}{C}\right); \quad \frac{O}{C} < 1 \tag{3.16}$$

For technical fuels, it is more convenient to introduce mass fractions. For wood, from Eq. (3.15),

$$\beta = \frac{1.0412 + 0.2160\left(\frac{Z_{H_2}}{Z_C}\right) + 0.2499\left(\frac{Z_{O_2}}{Z_C}\right)\left[1 + 0.7884\left(\frac{Z_{H_2}}{Z_C}\right)\right] + 0.0450\left(\frac{Z_{N_2}}{Z_C}\right)}{1 + 0.3035\left(\frac{Z_{O_2}}{Z_C}\right)} \quad \frac{O}{C} < 1 \tag{3.17}$$

For liquid technical fuels, from Eq. (3.16),

$$\beta = 1.041 + 0.1728\left(\frac{Z_{H_2}}{Z_C}\right) + 0.0432\left(\frac{Z_{O_2}}{Z_C}\right) + 0.2169\left(\frac{Z_{H_2}}{Z_C}\right) + \left[1 - 2.0628\left(\frac{Z_{H_2}}{Z_C}\right)\right] \tag{3.18}$$

where H/C, O/C, and N/C denote atomic ratios of the elements and Z_i represents mass fractions of the elements in an organic compound. For most technical fuels, the factor β is higher than 1, which means that the chemical exergy is higher than the corresponding LHV. The average β value for wood is 1.15, for hard coal, lignite, and coke is 1.06–1.17, for liquid hydrocarbon fuels is about 1.07, and for natural gas is 1.04. The correlation of the standard chemical exergy of several organic fuels to the lower and higher heating value is shown in Table 3.2, according to Szargut et al. [14].

Song et al. [16] proposed a simple method to estimate the specific chemical exergy of dry biomass employing the higher heating value (HHV):

$$ex_{ch} = 1.047\ HHV_{db} \tag{3.19}$$

In commercial simulators such as Aspen Plus or PRO/II, Eq. (3.19) can help perform the biomass-based system's exergy analysis. Although this simulation software can define and calculate HHV based on the proximate and ultimate analysis, it cannot calculate biomass's chemical exergy.

Table 3.2 The ratio of chemical exergy of organic fuels to the lower heating value (LHV) and higher heating value (HHV).

Fuel	ex_{ch}/LHV	ex_{ch}/HHV
Hard coal	1.09	1.03
Lignite	1.17	1.04
Coke	1.06	1.04
Wood	1.15	1.05
Liquid HC-fuel	1.07	0.99
Natural gas	1.04	0.94

3.3.3 Exergy efficiency

In the real process, exergy is consumed, where it differentiates with mass and energy balances. Exergy itself represents the system's useful work potential, and exergy destruction is related to irreversibilities within the system or during the process. Exergy destruction related to entropy generation is considered bad because any irreversibilities should usually be minimized. The overall exergy rate balance for control volumes at a steady state is illustrated in Fig. 3.9 for a single system. The same concept can be extended for an integrated complex system consisting of various processes.

In this study, the exergy analysis of a system is performed by modeling exergy balance as follows:

$$\sum_{i,\ in} \dot{m}_i\, ex_i + \sum_{q,\ in} \dot{E}_Q - \sum_{w,\ in} \dot{E}_W = \sum_{e,\ out} \dot{m}_e\, ex_e + \sum_{q,\ out} \dot{E}_q - \sum_{w,\ out} \dot{E}_w + \dot{I} \quad (3.20)$$

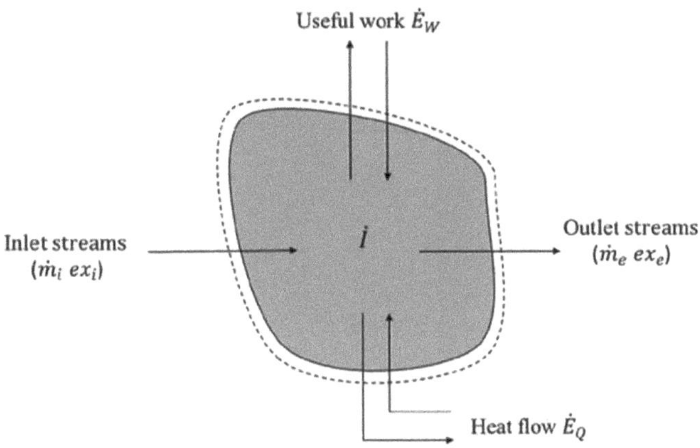

Figure 3.9 Overall exergy rate balance for steady flow process.

where, m, ex, $\dot{E}_Q$, $\dot{E}_W$ are the mass flow rate, exergy per unit of mass, thermal exergy rate, the exergy rate of a work transfer, respectively. $\dot{I}$ denotes the exergy rate loss due to real irreversible processes taking place inside the system. In the actual application, the heat flow or thermal exergy rate $\dot{E}_Q$ is sometimes the difficult one to estimate, especially if the exact temperature of heat transfer is unknown. It is worth noting that the sign convention used in this book for heat and work is that the heat flow from the surroundings to the system is taken as positive, and the work transferred from the system to the surroundings is also positive. The total exergy (ex) is the sum of physical exergy (ex_{ph}) and chemical exergy (ex_{ch}) and calculated as follows:

$$ex = ex_{ph} + ex_{ch} \tag{3.21}$$

$$ex_{ph} = (h - h_0) - T_0(s - s_0) \tag{3.22}$$

$$ex_{ch} = \sum_e x_i \left[ex_{ch,i} + RT_0 ln\left(x_i / \sum x_i \right) \right] \tag{3.23}$$

where T, h, s, x, and R are temperature, enthalpy, specific entropy, molar fraction, and the universal gas constant, respectively. By using the definition of temperature and pressure-based exergy, the following general expressions can be used:

$$ex_{ph}^T = (h(T,p) - h_0(T_0,p)) - T_0(s(T,p) - s_0(T_0,p)) \tag{3.24}$$

$$ex_{ph}^p = (h(T_0,p) - h_0(T_0,p_0)) - T_0(s(T_0,p) - s_0(T_0,p_0)) \tag{3.25}$$

For an ideal gas, enthalpy is only a function of temperature, while entropy is a simple function of both pressure and temperature,

$$h - h_0 = \int_{T_0}^{T} c_p dT, \tag{3.26}$$

$$s - s_0 = \int_{T_0}^{T} \frac{c_p}{T} - \int_{p_0}^{p} R \tag{3.27}$$

Considering the mathematical definitions for temperature based and pressure-based exergy, the following formulas can be arranged for an ideal gas case with constant specific heat capacities.

$$ex_{ph}^{T} = c_p\left[T - T_0\left(1 + \ln\frac{T}{T_0}\right)\right] \quad (3.28)$$

$$ex_{ph}^{p} = T_0.R.\ln\frac{p}{p_0} = \frac{k-1}{k}c_p.T_0.\ln\frac{p}{p_0} \quad (3.29)$$

From the equation, it is known that the exergy can be manipulated by changing temperature and pressure, and the two components can be traded off against each other within the total thermomechanical exergy. In cryogenic processes, a pressurized (gas) stream can be expanded to provide extra cooling and power. In this operation, pressure-based exergy is converted into temperature-based exergy [11]. The situation would be different when above ambient environment. A turbine or expander takes both temperature and pressure (reduced) to the near ambient conditions to generate the power. Above ambient conditions, the main purpose of using a turbine or an expander usually is to generate power while the main objective below ambient is to produce cooling with work as a byproduct.

Fig. 3.10 depicts an example of the exergy balance for an open steady flow system. Our discussion is more common in defining an exergy efficiency by counting only useful exergy produced (delivered) by the system [11]. The exergy inputs can be any matter streams, work, and heat. The product or exergy output can further be split into exergy of products and waste exergy, where the latter represents the streams directly released to the environment. In a steady flow process, the exergy rate entering the system is always greater than leaving the system.

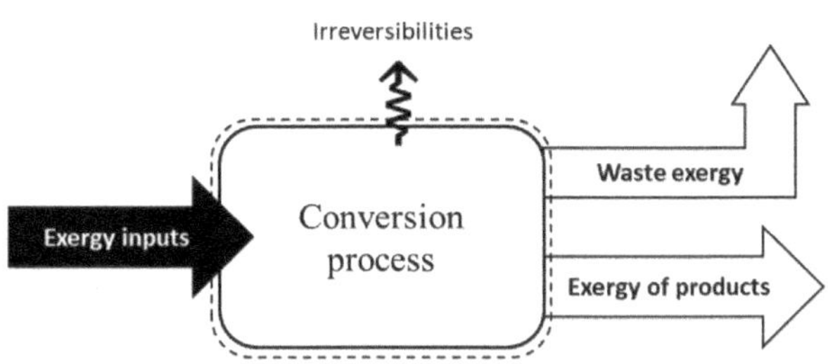

Figure 3.10 Exergy balance of a system for a steady flow process.

The difference between the exergy input and the exergy output of the system is called the exergy loss. Exergy loss can be divided into primary (irreversibilities) and secondary (waste exergy) loss. The primary exergy loss in a steady flow process, illustrated in Fig. 3.7, is due to the quality degradation of energy and materials and is listed in Eq. (3.18). The correlation of the primary exergy loss takes place inside the system and the entropy production rate $\left(\dot{\Pi}\right)$ can be written as:

$$\dot{I} = T_0 \dot{\Pi} \tag{3.30}$$

where T_0 denotes the environmental temperature. For the complex system, the total exergy losses are the sum of all primary exergy losses or destruction of each process components and can be calculated as:

$$\dot{I}_{total} = \sum_{k} \dot{I}_k \tag{3.31}$$

Primary exergy destruction is mostly inevitable exergy losses due to the progress of the process or reaction such as friction, unrestrained expansion of a fluid, heat transfer, mass transfer, chemical reactions that occur inside the system. For example, unavoidable exergy losses that rise with the conversion rate due to chemical reactions or uncontrolled chemical reactions produce heat in fuel cells. Avoidable exergy losses due to technical imperfections of processes can also lead to primary exergy destruction, such as high-temperature differences between streams in heat exchanger systems that result in significant heat loss. The thermodynamic imperfection of a process can be measured by the exergy destruction rate using exergy analysis to understand a system's performance. In various system components, exergy analysis can identify in which elements the most exergy destructions occur. In addition to primary destruction, secondary destruction is external losses that can also occur in a process. External loss is an exergy that is not practically utilized but rejected to the environment (waste exergy). Typical examples of external losses are exergy loss of flue gases from the power plants or exergy loss of wastes from various chemical processes.

Exergy efficiency is an important tool to evaluate system performance. Counting exergy efficiency can make more sense than counting energy efficiency since it would give efficiencies close to 100%. In this book, exergy efficiency is consistently defined as the ratio of products' exergy to the total exergy inputs. A typical exergy efficiency can then be expressed as follows:

$$\eta_{ex} = \frac{\text{Exergy of products}}{\text{Exergy inputs}} \tag{3.32}$$

Exergy efficiency formulations proposed in the literature have been summarized by Gourmelon et al. [17], as shown in Table 3.3 for further investigations and comparisons. The exergy efficiency concept has been used in many fields such as transportation, energy, agriculture, and chemical processes, to evaluate complex energy interactions taking place in a system. The exergy efficiency is considered a measure of the thermodynamic perfection of systems and processes. The use of exergy analysis can assess the cause of irreversibility or component efficiency. In more complex systems such as petrochemical, the exergy analysis is a powerful tool to design, evaluate, and optimize the overall system.

Rosen et al. [26] have reported the use of exergy to understand electrical power technologies' efficiencies and assist improvements. It is concluded that a high exergy loss or low exergy efficiency can occur in various devices. The high loss indicates that a significant margin for efficiency improvement exists in theory. Thus, exergy analysis has a significant role in investigating and improving various technologies and systems' performances. It also becomes a useful tool for engineers and scientists as well as policymakers in many fields. However, ingenuity and creativity are often required to implement the efficiency improvement. Also, by understanding the causes, locations, and magnitudes of the losses, efficiency-improvement efforts can be more focused.

Table 3.3 Several formulations for exergy efficiency according to several works of literature.

Name	Formula
Degree of perfection [18] Simple efficiency [19]	$\eta_{ex} = \frac{\sum \text{exergy output}}{\sum \text{exergy input}}$
Grassmann efficiency [20]	$\eta_{ex} = \frac{B_{useful\ output}}{B_{input}}$
Rational efficiency [21–23]	$\eta_{ex} = \frac{\text{Desired output}}{\text{Necessary input}}$
Exergetic efficiency [18]	$\eta_{ex} = \frac{\text{Useful exergy effect}}{\text{Driving exergy}}$
Fuel-product efficiency [24,25]	$\eta_{ex} = \frac{\text{Product exergy}}{\text{Fuel exergy}}$
	$\eta_{ex} = \frac{\text{Product exergy}}{\text{Fuel exergy} - \text{unavaidable irreversibility}}$

Furthermore, the interaction between exergy and economy named thermoeconomics or exergoeconomics has been widely studied in various cases. It combines exergy analysis and economic principles, leading us to understand how costs flow in a system and optimize its performance [27]. The process engineer is frequently demanded to reduce the input cost of a process by minimizing exergy loss due to system imperfections. The interrelations among exergy, economy, and environment have further developed to bring some fundamental changes in evaluation, design, and maintenance processes. This more comprehensive approach may have a positive impact on sustainable development and environmental protection [28].

3.4 Exergy analysis of biomass conversion process

The exergy analysis technique estimates the process's efficiency and determines the energy quality and usefulness [29]. Exergy analysis allows us to specify the system's maximum performance and the sources of the irreversibilities and has been used to evaluate different systems' performance. As discussed in Chapter 2, biochemical and thermochemical processes are two main methods to convert biomass into useful thermal energy/heat, power, or fuels. The thermochemical process can be divided into combustion, gasification, pyrolysis, liquefaction, and thermochemical cycle. The desired form of energy and biomass types is two important factors determining the conversion process [30]. Direct combustion of biomass is the traditional way of using biomass. Theoretically, it is possible to burn any biomass. However, the combustion process is feasible for the biomasses with less than 50% moisture content. Gasification is believed to achieve a higher conversion rate in comparison with other thermochemical methods.

Exergy analysis of biomass thermochemical conversion processes has been widely studied and reported earlier, such as gasification [31–33], combustion [34], pyrolysis [35], including liquefaction for bio-oil production [36]. Depending on the evaluated system desired, the exergy analysis of a biomass conversion process can also cover pretreatment and fuel postprocessing. For example, David and Zhang [23] have modeled and evaluated biomass gasification with a char-catalytic tar reforming system to understand the exergy and energy performance. They developed a model based on a combination of equilibrium thermodynamics and chemical kinetics. The biomass gasification process was simulated with a thermodynamic equilibrium model available in the literature. It was followed by a kinetics model, which consists of reforming the tar with residual char catalysts. The process performance was assessed based on the energy and exergy efficiencies.

Since biomass gasification is a complex process both from the experimental (costly and time-consuming) theoretical perspectives, model developments can be useful engineering tools to predict gasification behavior. The available models for the gasification process can be divided into thermodynamic equilibrium models, kinetic rate models, and neural networks models [37]. Thermodynamic equilibrium models are universal as they can be applied for a wide range of conditions. It is a useful tool for preliminary comparison and process studies on the most important fuel and process parameters' influence. The thermodynamic equilibrium approach has the advantage of being independent of gasifier design, but it cannot give highly accurate results for all cases. In that case, the kinetic rate models are more restricted to specific gasifiers and process conditions, computationally more intensive but give more accurate and detailed results. Various models are also developed using the simulator Aspen Plus, and they can include thermodynamic and kinetic model elements. Aspen Plus is a software suite used to perform simulations of complex physical, chemical, and biological processes [5]. Simulation of the process usually involves three main steps:

- A flowsheet consisting of the elementary blocks of unit operations is developed.
- The chemical components and specification of conditions of the process streams (flow rates, temperature, pressure, and compositions) are defined.
- The operating conditions for unit operations are specified.

There are two approaches adopted for modeling the process of biomass gasification, i.e., equilibrium and kinetic. The equilibrium approach, classified as stoichiometric and nonstoichiometric, is applied by assuming that the reactants are ideally mixed and react with each other in an infinite period [38]. However, the stoichiometric approach may cause some errors due to a lack of available information to predict complex processes. In this case, a nonstoichiometric modeling approach based on minimization of the Gibbs free energy can be applied to handle a complex process. In the simulator, the gasification process based on the nonstoichiometric method is divided into two main steps, decomposition and subsequent chemical reaction. In the Aspen Plus simulator, the decomposition and subsequent gasification reactor are simulated as the RYield and RGibbs blocks, respectively, as shown in Fig. 3.11.

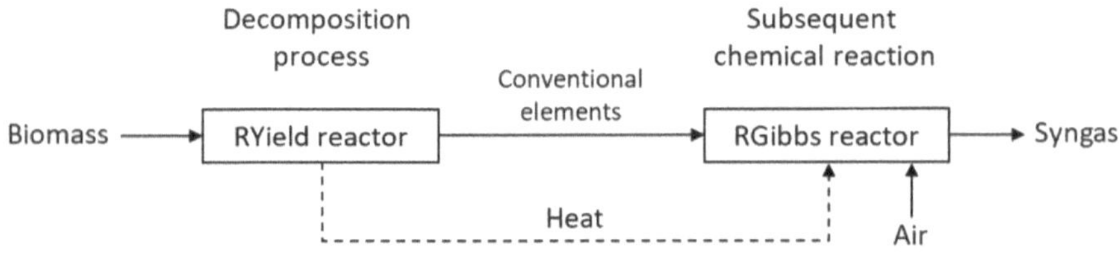

Figure 3.11 Aspen Plus model air-blown gasifier.

Since biomass is defined as a nonconventional solid, it cannot be reacted directly with other reactants. An RYield block with a FORTRAN subroutine is then used to decompose biomass solid into H_2O, ash, and elements, such as O_2 H_2, and carbon. During biomass decomposition, it is assumed that the devolatilization products' yield and composition can be derived from the ultimate and proximate analysis of the biomass. Therefore, the information about the proximate and ultimate analysis should be provided earlier. The decomposed components are then introduced into an RGibbs block to perform the subsequent gasification chemical reactions. By adopting this method, it is unnecessary to specify complex chemical reactions in the gasification process. The heat of reaction for biomass conversion is equal to the sum of the heat of the RYield (biomass decomposition) block and RGibbs (gasification reaction) block.

Kinetic modeling, on the other hand, is generally applied for predicting the syngas yield and the composition after a finite time. It results in a more realistic approach concerning the real operating conditions and gasification configuration, particularly for lower gasification temperatures and shorter residence times. However, kinetic models are computationally more intensive and subjected to the limited availability of experimental rate expressions. Various chemical and physical phenomena can be included in kinetic models, such as kinetics of homogeneous and heterogeneous chemical reactions, heat and mass transfer between solid and gaseous phase, and transport of produced volatiles. Most of the kinetic models are developed for fixed-bed, particularly downdraft gasifiers [5]. In this gasifier type, the model is usually divided into different parts corresponding to specific zones (drying, pyrolysis, combustion, and reduction). The development of kinetic models for biomass gasification is still the subject of intensive investigations. From this perspective, the application of kinetic rate models for commercial biomass gasifiers' design is rather limited.

The simplified gasification system is illustrated in Fig. 3.12 to evaluate its exergy performance. The solid biomass and the gasifying agent (air, oxygen,

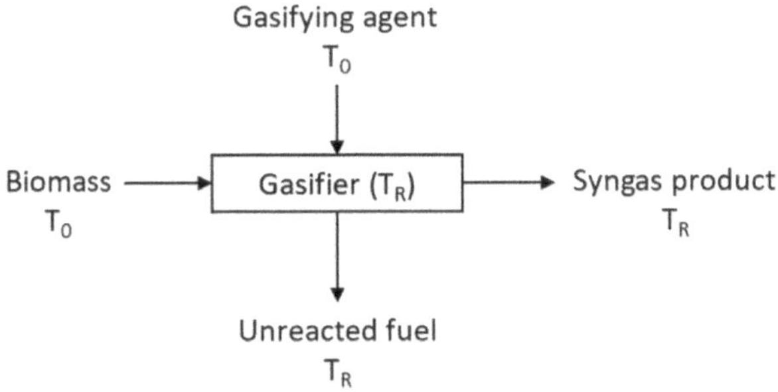

Figure 3.12 Schematic diagram of the gasification process.

and/or steam) are introduced to the gasifier at the environmental temperature T_0. In this case, the gasifying agent at atmospheric temperature-pressure conditions contains no exergy and then negligible for exergy calculations. The gaseous product (syngas) and unreacted fuel leave the gasifier at the reactor temperature T_R. The process is assumed that the gasifier operates as an adiabatic and pseudo-homogeneous reactor. Gasification entails partial oxidation of the feedstock, so biomass's chemical energy is converted into a syngas product's chemical and thermal energy. Cold gas efficiency as a traditional gasification parameter can be calculated as the ratio between cooled syngas' energy to the input corresponding biomass energy. The fuel and the gas's chemical energy values to calculate the cold gas efficiency formula can be in the form of either HHV or LHV. However, the exergy efficiency may be defined as the ratio between the sum of chemical and physical exergy of syngas product and biomass fuel,

$$\eta_{ex} = \frac{\dot{E}_{syngas}}{\dot{E}_{biomass} + \dot{E}_{gasifying\ agent}} \tag{3.33}$$

3.5 Process modeling and exergy efficiency improvement

Currently, the design, operation, and control of the biomass conversion processes are the subject of investigation efforts for many researchers. Due to complexity, modeling and simulation are used to address the integration of process technologies in a system to enhance the understanding of complex unit operations. Modeling and simulation approaches (or combined with experiments) have been used to save time and money since they reduce the need for expensive and complicated large-scale experiments [39].

Process modeling for the biomass conversion system is a technique performed to understand and describe the integrated process from biomass into other forms of energy and products. It is developed by improving the connection of material and energy flows between the current and the future state of a process. Process modeling sometimes consists of not only process integration but also optimization. A diagram of the integrated process representing how to convert biomass into hydrogen involving gasification and steam reforming is an example of the process modeling result. It can further be a very useful technique to optimize biomass energy conversions.

A comprehensive list of software tools for the modeling and simulation of material and energy balances of chemical/energy processing plants has been listed by Klemes [40], such as Aspen Plus, Aspen HYSYS, UniSim Design, gPROMS, CHEMCAD, PRO/II, including general-purpose optimization systems (GAMS, MIPSYN). Performing the modeling, process integration, and optimization problems that are very complex requires suitable tools that provide fast and accurate solutions via a user-friendly interface. Those software tools have been widely used, which has helped small/medium scales or even process industry companies achieve their operational goals.

Example

Fig. E3.1 shows a conceptual diagram of the overall integrated gasification combined cycle. The integrated system mainly comprises a gasifier, heat recovery system, and power generation. Initially, the concentrated fuel and gasifying agent is introduced to the gasifier for syngas conversion. The choice of the gasifying agent is also motivated by the desired composition of an end product. After complex reactions occur inside the gasifier, the product gas (syngas) and the unconverted fuel leave the reactor at the high gasification temperature. Since the hot syngas contains a significant amount of exergy with high temperature, heat recovery is employed to generate steam for the subsequent gasification process. In this case, steam at atmospheric pressure is used optionally as a gasifying agent.

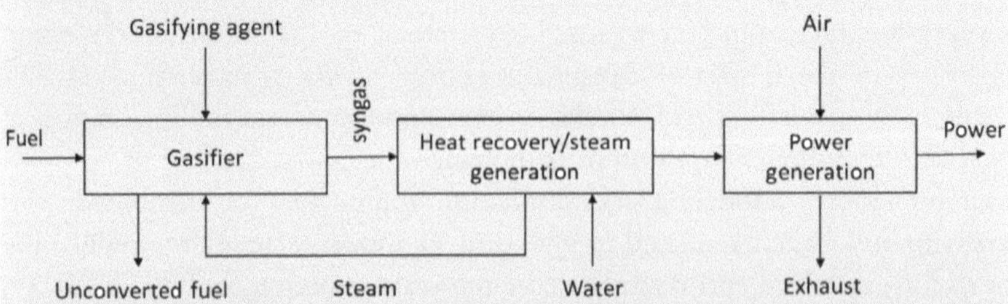

Figure E3.1 Simplified diagram of integrated gasification combined cycle (IGCC).

Example—cont'd

The lower temperature syngas is used as fuel for combustion, creating a high-temperature pressurized gas to rotate the gas turbine. As the temperature of the gas turbine's flue gas is still high, the rest of the heat will be utilized to generate steam in the heat recovery steam generator (HRSG), which is used to rotate the steam turbine to generate the electricity. The exhaust gas is released into the environment and constitutes some portion of the system's total exergy loss.

The overall exergy balance for the IGCC shown in Fig. E3.1 can be expressed as:

$$\dot{E}_{fuel} + \dot{E}_{gasifying\ agent} = \dot{E}_{unconverted\ fuel} + \dot{E}_{exhaust} + \dot{E}_{power} - \dot{E}_{compressor\ work}$$

Where $\dot{E}_{fuel}$ is the exergy rate of the incoming solid fuel, $\dot{E}_{unconverted\ fuel}$ is the exergy rate of the unconverted solid fuel, $\dot{E}_{gasifying\ agent}$ is the exergy rate of the gasifying agent (air), $\dot{E}_{exhaust}$ is the exergy rate of the exhaust gas, $\dot{E}_{power}$ is the exergy rate of the produced electricity, and $\dot{E}_{compressor\ work}$ is the exergy rate of work for air compression in power generation. For most energy conversion technologies, the process efficiency is defined as the desired output ratio to the process input. Assuming the power is the only desired output of the IGCC process, the overall exergy efficiency η_{ex},

$$\eta_{ex} = \frac{\dot{E}_{power}}{\dot{E}_{fuel} + \dot{E}_{gasifying\ agent}}$$

According to the equation, the unconverted fuel and exhaust gas are not classified as a product but as an external exergy loss. However, unconverted fuel can also be a useful product if recovered as a product or reintroduced to the gasifier.

There are two engineering design problems in integrated processes. The first is the problem of unit operation design and the second is designing total systems. Our discussion in this book mostly addresses the latter issue in the process flowsheet's particular design to improve exergy efficiency. Earlier, a union diagram has been presented in Chapter 1 to show how operations within a process or several processes are combined to improve energy efficiency and avoid harmful emissions. To understand practical rules for improving exergy recovery, the most well-accepted representations for chemical process design are presented in Fig. 3.13, developed by Klemes et al. [40]. The reaction layer is the core of the whole process, and its design influences the separation structures in the second layer of

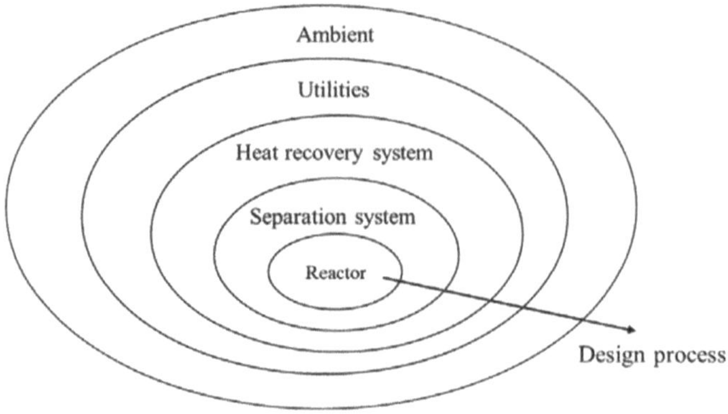

Figure 3.13 The conventional onion diagrams.

the union. The reactor and separator structures determine the overall heat recovery system. Thus, the heat recovery system is designed next. The fourth layer's energy utility system is designed to provide additional heating and cooling requirements that cannot be satisfied through the heat recovery system [41]. The ambient system at the outer layer is provided for handling the various emissions/effluents emitted from the process before final discharge to the environment.

In the energy process design, exergy analysis should be adopted to reduce the thermodynamic imperfection. Exergy analysis reveals the relative magnitude and nature of the irreversibilities of existing processes. In the new plant designs, understanding the principle of exergy analysis can significantly improve the process efficiency. However, the exergy efficiency (thermodynamic effectiveness) of a process should also be considered together with other design criteria, such as economics, reliability, safety, and operability, often conflicting [5]. There are several rules and guidelines for improving exergy efficiency reported by Ref. [5] that have been initially arranged by Szargut [14], Sama [42], Leites [42], and Kenney [43]; see Table 3.4.

The practical rules for exergy efficiency improvement can be classified into two categories: general and specific. In general rules, excessively large or excessively small driving forces such as temperature should be avoided. For example, the temperature differences between the hot and cold fluid within a heat exchanger should not be too small and not too large. If the temperature differences are too small, a large heat transfer area would be required, and basically, it is costly. If too large, the efficiency would be reduced as the entropy production increases; thus, an appropriate

Table 3.4 Rules and guidelines for improving exergy efficiency.

	Part of the process	Rules
Specific layer	Reactor	• Perform exothermic reactions at the highest possible temperature, whereas the endothermic at the lowest possible • A couple of exothermic and endothermic reactions in the same reactor • Avoid using diluents in both exothermic and endothermic reactions
	Separation system	• Perform separations at conditions that maximize separation efficiency. • Maximize heat reuse (evaporation, distillation) by using multiple effects systems and heat pumps. • Integrate the separation process into a chemical reactor
	Heat recovery system	• Match streams for heat exchange with similar heat capacities • Adopt pinch analysis, match streams for a heat exchange that the final temperature of one is similar to the initial temperature of the other • Do not discard the high-temperature heat to the ambient or cooling water • Avoid the use of intermediate heat transfer fluids
	Utilities	• Place fans and compressors in the coolest place • Expanders could be used to recover wasted energy instead of throttling valves • Avoid compression of the previously compressed steam of gas medium
	General guidelines	• Avoid excessive thermodynamic driving forces • Perform counter-current processes since they are generally thermodynamically more efficient than co-current • Accept exergy losses only if any economic justification • Better to avoid the elongation of thermodynamics process chains • Minimize the mixing of streams with different temperature, pressure, or chemical composition

temperature difference should be proportional. Another critical point is, adopting counter-current processes are generally more preferable to cocurrent. The counter-flow heat exchanger ensures a higher temperature increase of the heated stream at the same heat-transfer area than the cocurrent heat exchanger.

Furthermore, exergy losses can be accepted if any economic justification; for example, exergy losses due to irreversible heat transfer cannot be eliminated from a heat exchanger because the required heat transfer area would be infinitely large. Also, try to avoid the elongation of thermodynamics process chains because every new process introduces new exergy losses and additional investment. Lastly, minimizing the mixing of streams with different temperature, pressure, or chemical composition can avoid significant exergy losses. The mixing of streams with different temperatures is equivalent to an irreversible heat transfer.

In the reactors (first layer), a high-temperature operation of exothermic reactions can increase the produced heat or higher exergy value. Otherwise, a low-temperature process of endothermic reactions requires a lower energy supply. Hence, it is preferable to carry out either exothermic reactions or endothermic reactions at the highest and lowest possible temperatures. Coupling exothermic and endothermic reactions in the same reactor can also reduce entropy generation due to longer chains. In the separation process, the operations driven by heat, such as evaporation, can be performed by maximizing the heat reuse employing multiple effects systems. By integrating the separation process into a chemical reactor, exergy losses can be minimized. There are some guidelines in the utility layer by placing fans and compressors in the coolest place, adopting expanders to recover wasted energy instead of throttling valves, and avoiding compression of the previously compressed steam of gas medium.

The heat recovery system's guidelines emphasize that maintaining the uniform temperature differences between streams in a heat exchanger is essential. In a complex system involving many processes, adopting pinch analysis can match heat exchange streams, thus reducing heat losses. Pinch Analysis principles are employed to identify the energy-saving potential for the process and subsequently to aid the design of the heat exchanger systems. This approach's basic idea is to correctly perform process integration technology of system consisting of heat recovery, heating, cooling to reduce energy consumption. However, the studies reported that pinch analysis mostly focused on heat exchange and heat recovery networks instead of manipulating the system to make a more efficient system. Pinch analysis can be performed after analyzing all processes involved and completing a mass and exergy balance assessment of the reaction. It has successfully been applied at various industry levels to improve heat recovery and address utility systems. The main limitation of the heat pinch technology is that it only considers temperature as a quality parameter.

Example

An example of the heat exchanger process is presented with a "black-box" chemical reaction, as seen in Fig. E3.2. The feed (liquid) is introduced to the reactor and initially needs to be heated from ambient temperature (assumed to be 25°C) to the operating temperature of 250°C. Any flow that requires heating or cooling, but does not change in composition, is defined as a stream. The liquid feed, which needs to be heated, is called a cold stream. Conversely, a hot product from the separation system is called a hot stream. The hot stream needs to be cooled down to a lower temperature. Some heat from the hot stream can then be recovered to heat the cold stream in a heat exchanger (HX). However, based on thermodynamics' second law, it is impossible to use a hot stream at 200°C to heat a cold stream at 250°C. A heater (H) can be utilized to satisfy the remaining duties with less energy requirement. Overall, energy consumption for the heater can be reduced. If the data for simple two streams are known (such as mass flow rate and specific heat capacity), the amount of heat that can be recovered and energy consumption for the heater can also be determined.

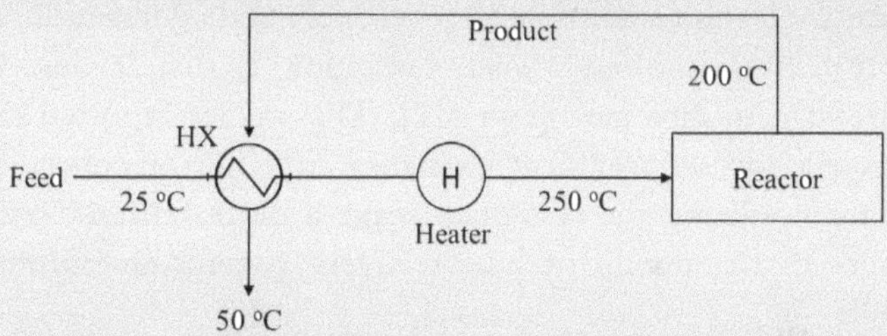

Figure E3.2 Simplified diagram of the heat recovery process

3.5.1 Pinch analysis

Several papers and books have been written to discuss and address the pinch analysis of the system [12,44,45]. This approach's basic idea is to correctly perform process integration technology of a system that consists of heat recovery, heating, and cooling to reduce energy consumption. In the conventional onion diagram (Fig. 3.8), pinch analysis is performed in layer three as part of the heat recovery system. Using the pinch analysis has advantages related to capital and operational costs, saving up to

15%–45% [46]. Pinch analysis has been widely applied for the biomass conversion process. For example, Fujimoto et al. [47] performed and applied pinch analysis for the bioethanol production process from lignocellulosic biomass. Their findings indicated that using heat released during the process and surplus power generated, pinch analysis could improve energy efficiency with an energy saving of 38%.

Among the advantages, pinch analysis has the ability to establish performance targets ahead of design based only on information about the change in the thermodynamic state for process streams. These targets include thermal energy (external heating and cooling), mechanical energy (power or shaft work), number of heat exchangers, and total heat transfer area [48]. Pinch analysis is performed after analyzing all processes involved and completing a mass and exergy balance assessment of the reaction. Then, heating and cooling requirements can be listed by understanding the exergy quality of each stream. Next, the composite curves (T-H diagram) and the heat exchanger network can be constructed and developed to show heating and cooling processes' profiles. Pinch analysis is based on straightforward thermodynamics and uses it practically. It has successfully been applied at various industry levels to improve heat recovery and address utility systems. The heat pinch technology's main limitation is that it only considers temperature as a quality parameter [12]. The studies reported that pinch analysis mostly focused on heat exchange and heat recovery networks instead of manipulating the system to make a more efficient system. The typical stages in the design of heat recovery systems are summarized as follows [49]:

- **Data extraction.** Based on the heat and mass balance of processes, the heating and cooling requirements are collected. In this stage, the most important is to ensure the consistency of heat and mass balance than can be done by process simulators. However, data quality depends on whether the simulation model represents the actual process.
- **Performance Targets.** Based on stream data from the data extraction, the best performance ahead of design is established. Typical targets for Heat Exchanger Networks include minimum external heating ($Q_{H,min}$) and cooling ($Q_{C,min}$) demands, the minimum number of heat exchangers (U_{min}), and the minimum total heat transfer area (A_{min}).
- **Process modifications.** It refers to the consideration of making changes in the basic process (reactor system, separation system, recycle system, etc.). This step can reduce energy targets or give simpler and cheaper networks.

- **Network design.** It refers to establishing a network of heat exchangers with a Maximum Energy Recovery (MER) design. A design that has fewer heat exchangers and less stream splitting as possible is usually favorable. Separate networks are established above and below pinch based on firm rules for matching hot and cold process streams in heat exchangers.
- **Design evolution.** It is performed by removing small to refine heat exchanger network streams. This stage is needed to reduce network complexity as well as cost.
- **Process simulation.** This final stage is required to test the Heat Exchanger Network's feasibility that has been designed and optimized in the previous steps. This stage may also involve switching from simplified models of pure counter-current heat exchangers to more practical and realistic design configurations. This stage may also include changing from simplified models of heat exchangers to more practical and realistic design configurations.

3.5.1.1 The temperature–enthalpy diagram and composite curves

The temperature–enthalpy diagram (T-H diagram) is a useful method to visualize the composite curve for hot or cold streams and to represent heat exchange. It can be constructed from any pure substance depending on the process involved. Assuming we have a process involving one heat exchanger, one cooling, and one heating utility, the alternative representations of the heat exchanger process can be shown in Fig. 3.9A. The heating utility is needed to raise the cold stream to the target final temperature due to hot stream limitation. Conversely, heat available in the hot stream has to be rejected to cooling water. The heat recovery, heating, and cooling duties are represented by Q, Q_H, and Q_C, respectively. Fig. 3.14B shows an example of the hot and cold streams plotted on the T/H diagram. $T_{H,in}$ and $T_{H,out}$ represent the hot stream entering and leaving the system, respectively. The hot stream is represented by the line with the ar pointing to the left and the cold stream vice versa.

A single composite for multiple hot streams and a single composite for multiple cold streams can be produced in the T/H diagrams to provide a counter-current picture of heat transfer. They can be combined and used to illustrate the minimum energy target for the process. The composite curves can be obtained by lumping all the heat from different streams at the same temperature interval, as shown in Fig. 3.15. However, we lose information on the vertical arrangement of heat transfer between streams by

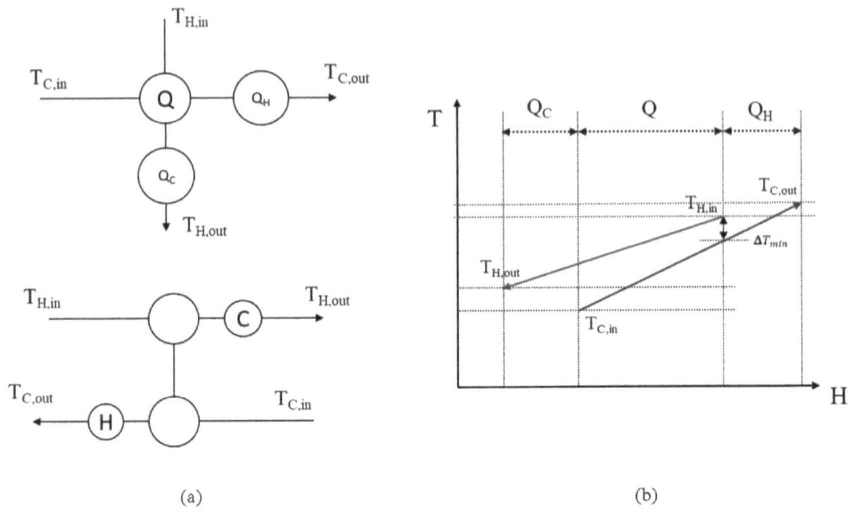

Figure 3.14 (A) The two alternative representations of the heat exchanger and (B) T/H diagram.

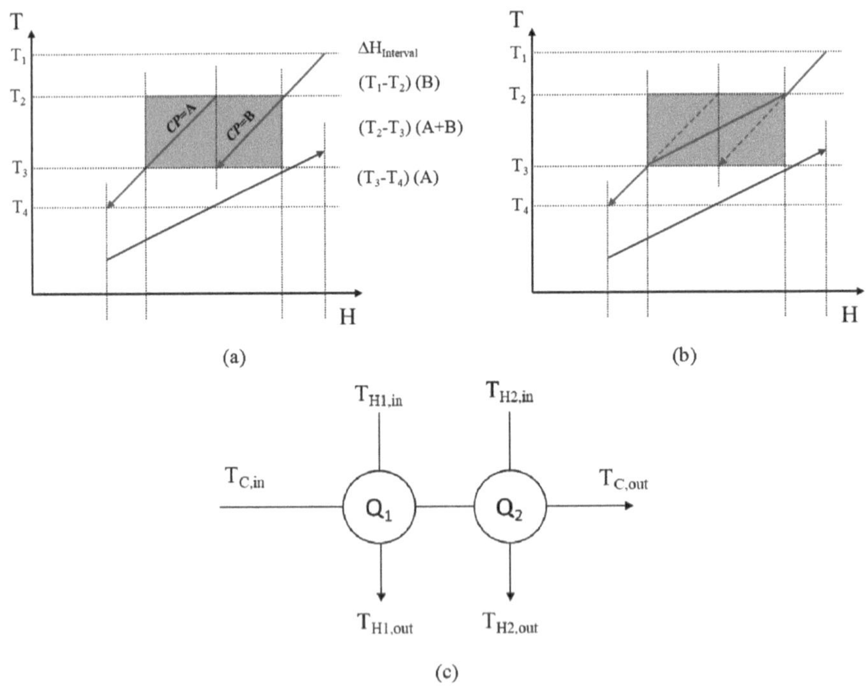

Figure 3.15 Formation of the composite curve.

constructing a composite curve [50]. Two hot streams are plotted separately in Fig. 3.15A, with their supply and target temperatures defining a series of "interval" temperatures T_1-T_4. Between T_1 and T_2, only stream B exists, and so the heat available in this interval is given by (T_1-T_2) (B). Since streams A and B exist between T_2 and T_3, the heat available in this interval is (T_2-T_3) (A + B). By lumping all the heat from the hot streams, it is replotted against the interval temperatures shown in Fig. 3.15B. The T-H plot resulted in a single curve representing all the hot streams, known as the hot composite curve. However, in this example, only one cold stream is involved. The same way can be done if several cold streams are available in the process. The overlap between the composite curves represents the maximum heat recovery possible within the process [44].

The value of ΔT_{min} is the temperature difference between hot and cold streams at the cold end of the exchanger. In general, ΔT_{min} occurs at only one point of the closest approach, called the pinch, and typically divides the temperature range into two regions (not just at one end), as seen in Fig. 3.16. The heating utility can be used only above the pinch and the cooling utility only below it. If we choose a larger value of ΔT_{min}, the heating utility will require higher energy consumption. The higher heating utility load will result in a higher cooling utility load. These fundamental facts are important in designing a heat exchanger system practically.

3.5.1.2 The use of composite curves to determine the energy targets

The minimum energy target or minimum energy consumption for the process can be determined by moving the cold composite curve toward the

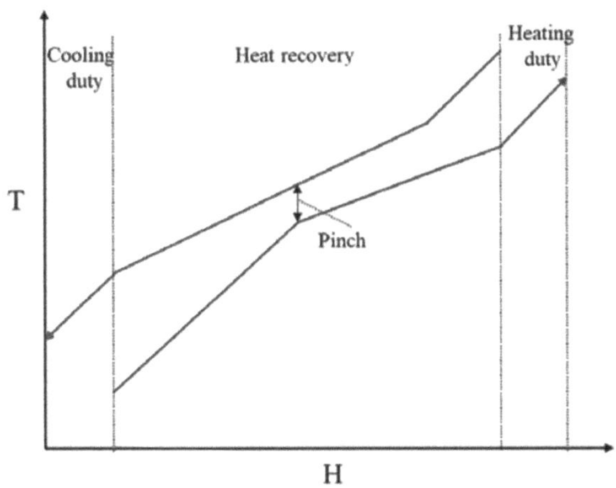

Figure 3.16 Composite curves of multiple streams.

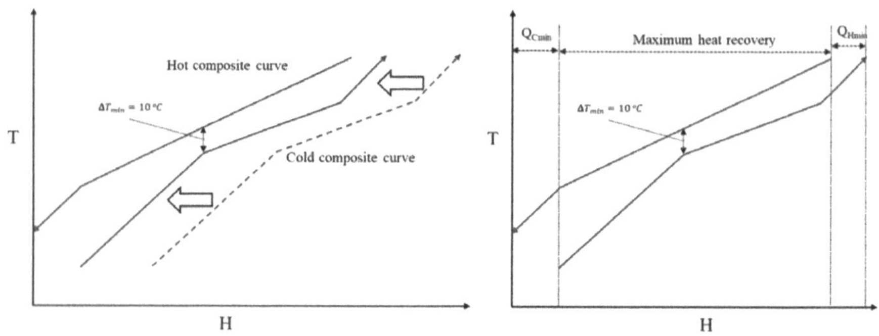

Figure 3.17 Using the composite curves to determine the energy targets.

hot composite curve, as shown in Fig. 3.17. Moving the cold composite curve in a horizontal direction does not change the stream data. The curves' closest approach is defined by the minimum allowable temperature difference, ΔT_{min} (this example uses 10°C as an example). The only modification needed is to ensure that cold and hot composite curves are at least ΔT_{min} apart within any interval. This value determines the minimum temperature difference that will be accepted in a heat exchanger. The minimum energy target means the minimum of both heating and cooling duties can be achieved with maximum heat recovery. We can set targets for this minimum energy consumption before designing any heat exchanger network using pinch analysis. This approach allows us to quickly identify the scope for energy savings at an early stage of the investigation. This single benefit is likely to be the biggest strength of the approach [51].

Practically, if composite curves of hot and cold streams are almost parallel, a higher value of ΔT_{min} will be chosen. The temperature difference between cold and hot streams is close to the ΔT_{min} value when the composite curves are almost parallel in any heat exchanger process. In this case, a small ΔT_{min} would result in a high heat exchange area for all heat exchangers (not only for the ones that transfer heat between streams close to the pinch point). It means high investment costs are required. In systems where fouling readily occurs or where heat transfer coefficients are low, typical ΔT_{min} values of 30–40°C are used. For chemical processes and where utilities are used for heat transfer, ΔT_{min} values are typically in the range of 10–20°C. For low-temperature processes using refrigeration, lower ΔT_{min} values (3–5°C) are used to minimize the refrigeration systems' expensive power demands.

Fig. 3.18 shows clearly the difference between pinch analysis and the conventional approach in designing the heat recovery system. In the conventional designing approach, the core of the process is designed with

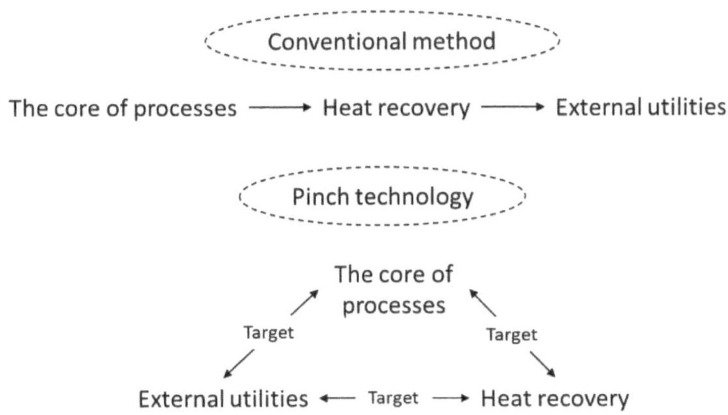

Figure 3.18 The difference between the conventional design and the pinch technology method. *(Adapted from Rokni M. Introduction to pinch technology division of energy section introduction to pinch technology. Technical University of Denmark; 2016.)*

fixed flow rates and temperatures so that the heat and mass balance of the process is satisfied. Afterward, the heat recovery systems are designed, and the remaining duties are completed using external utilities. In the pinch technology method, heat recovery systems are simultaneously designed together with the main process and external utilities. This approach has advantages over the conventional methods due to the ability to set capital cost and energy recovery targets for a subprocess or entire process ahead of design. Also, the energy savings and investment requirements are known in advance. However, although the pinch technology method has allowed many processes to become more efficient in energy consumption, it does not mean that the designed process is less-cost. Comprehensive evaluations are needed considering the interrelations among exergy efficiency itself, economy, and environment for a sustainable system.

3.6 Enhanced process integration: new approach

Throughout the past three decades, several onion diagrams have been proposed to improve the design process. This book is limited to discussing the second and third layers to minimize chemical processes' energy consumption. The pinch analysis discussed in the previous section is an optimization of heat recovery systems. In this section, enhanced process integration is proposed, and the onion diagram depicted in Fig. 3.13 can now take a revised form, as seen in Fig. 3.19. As shown, the material-recovery system and exergy elevation have been added in the second and third layers, respectively. In the context of process integration,

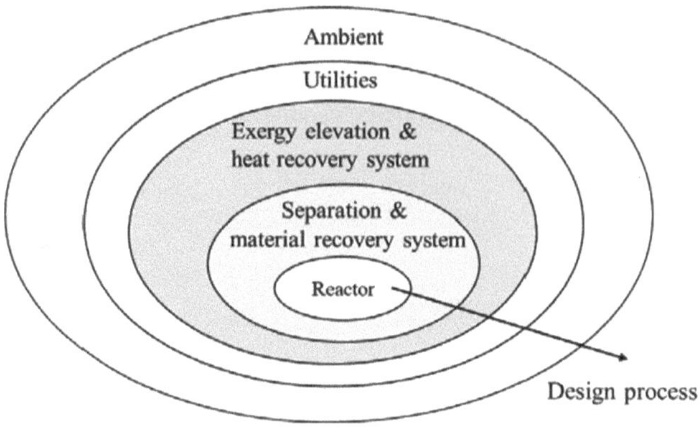

Figure 3.19 Onion diagram developed in this study.

the layer of the material-recovery system involves various material-recovery strategies, such as direct reuse/recycle and material regeneration. Furthermore, a holistic approach to process design, retrofitting, and operation should be considered, emphasizing the unity of the process to achieve an optimum result [41,53].

3.6.1 Separation and material recovery system

A material recovery strategy is introduced to reduce fresh resources and discharge waste materials through reuse, regeneration, and recycling. Several techniques for material recovery are now widely accepted. The term "reuse" means that the output material from a resource-consuming unit is sent to other operations and does not reenter processes from where it was emitted. The recycling scheme permits the effluent to reenter the processes wherein it was generated [41]. In regeneration schemes, output material is partially treated before reuse or recycle. The scheme of direct reuse and other material-recovery strategies can be seen in Fig. 3.20.

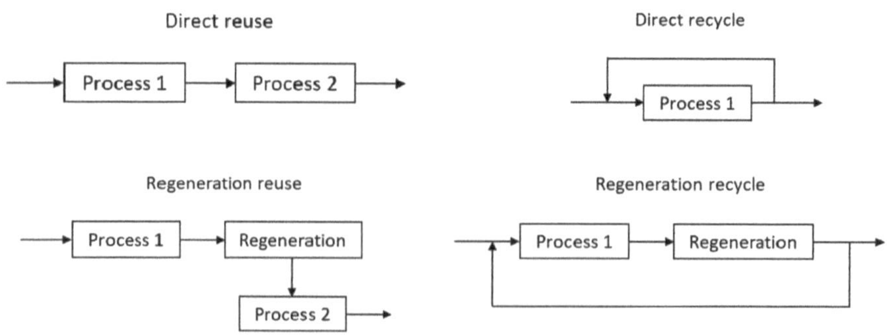

Figure 3.20 Various material-recovery strategies, including direct reuse, direct recycling, regeneration reuse, and regeneration recycling.

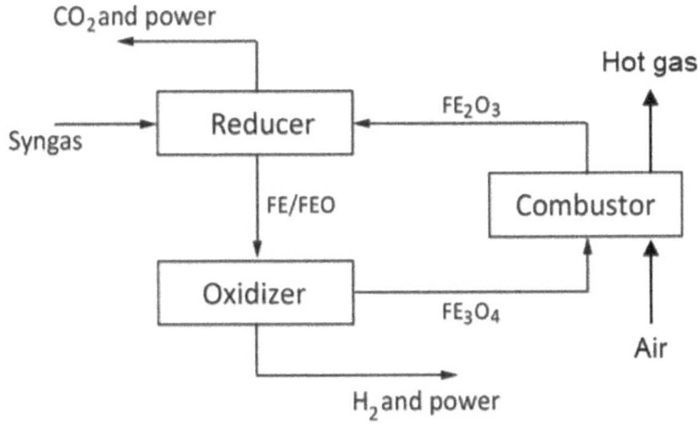

Figure 3.21 The recycling of iron-based oxygen carrier.

As in pinch analysis, data extraction is usually the most crucial step before adopting material-recovery strategies. The reuse, recycle, or regeneration techniques can be implemented and combined within an integrated plant implementing a process integration.

Fig. 3.21 shows an example of recycling the metal oxide in syngas chemical looping (SCL) technology. In the SCL module, coproduction of H_2 and power from syngas is conducted by cyclic treatment of iron oxide-based looping medium with syngas and steam. The SCL process consists mainly of reduction, oxidation, and combustion in interconnected reactors to produce pure H_2, CO_2, and power. Initially, syngas is introduced to the reducer and reacted with O_2 carried by Fe_2O_3. In the oxidation stage, the reduced particles react with steam to produce an H_2-rich gas stream, flowing out together with the remaining steam. Once the steam is condensed out from the gas stream, very high purity of the H_2 stream can be obtained. In the combustor, Fe_3O_4 produced in the oxidation reactor is reacted with O_2 to convert it back to Fe_2O_3.

3.6.2 Biomass drying based on heat circulation technology through exergy elevation and heat pairing

In a thermal drying system, a process for dehydration based on moisture evaporation by heating, exergy losses due to large irreversibility become a predominant factor causing a huge energy consumption. Imbalance amounts of heat exchange create the exergy loss due to nonoptimal heat pairing and unrecoverable heat brought about by the limitation of exergy rate differences among the streams. Hence, only a small part of the energy could be recovered, and a large exergy loss is experienced.

Therefore, the idea of innovative and effective biomass drying is developed by employing heat circulation technology to reduce the required drying energy. Unlike heat cascade utilization, heat circulation technology is developed based on exergy recovery [54]. To achieve this exergy recovery, first, the process stream's exergy rate is raised employing work such as compression to provide a minimum temperature difference for heat exchange (see Fig. 3.15B). The exergy loss could be minimized by converting combined thermal energy with a lower exergy rate and work into thermal energy with a higher exergy rate. Second, to achieve a balanced heat exchange, an effective heat pairing is conducted for sensible and latent heat. It could be performed through each stream's optimal pairing based on each stream's pattern revealed in temperature—enthalpy diagrams. As a result, all of the energy involved in drying can be recovered and utilized as the energy source for the subsequent drying step.

Fig. 3.22A shows the basic philosophy of the heat circulation technology employing exergy elevation applied in a drying process. A hot stream can be created through exergy elevation (process stream) by

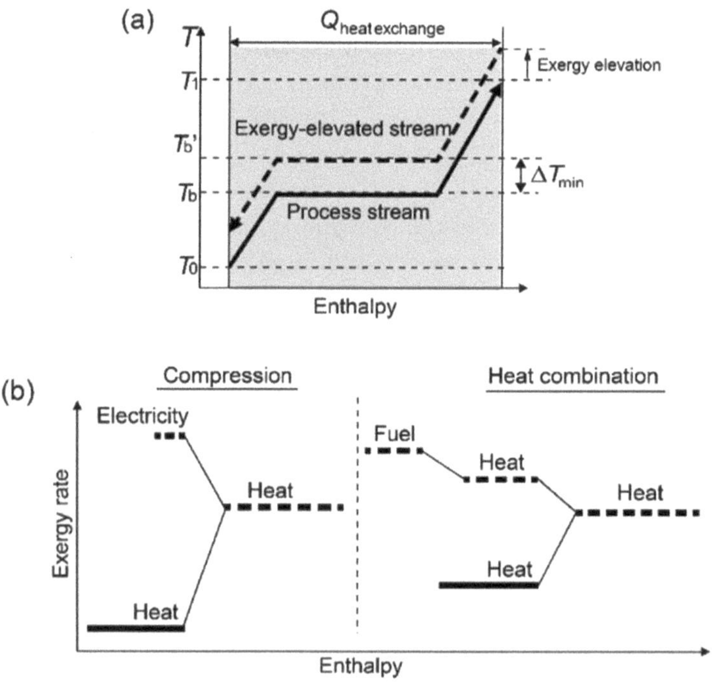

Figure 3.22 Exergy recovery principle: (A) exergy elevation of a stream and heat coupling between the hot stream and process streams, (B) exergy elevation through compression and heat combination.

compression of the evaporated water. Hence, the exergy rate of the evaporated water increases following the compression. This method can realize a self-heat exchange and minimize exergy destruction. Also, optimal heat exchange depends on heat pairing and the type of heat exchanger suitable for each process. Process integration is then employed in a bigger system to effectively combine the involved modules and efficiently utilize the heat that cannot be recovered anymore in each module to the other modules. The combination of material and exergy recovery (second and third layers) leads to a more optimized system and a larger amount of heat that can be recovered. Hence, the minimization of exergy destruction throughout the integrated system can be achieved, resulting in excellent total energy efficiency.

Example of algae drying

A steam tube rotary dryer with an internal heat exchanger is selected as the main algae dryer performing the drying process. The steam tube rotary dryer's advantages are the possibilities of a large heat transfer area with high thermal efficiency, high handling ability and suitability for continuous operation, excellent drying control, simple operation, and ease of use. In the steam tube rotary dryer, heat is exchanged conductively; hence, it is considered as indirect heating because the heat source is isolated from the material to be dried. The heating tubes are filled with compressed steam and arranged in a concentric circle inside the dryer, supplying the heat required for drying. The heating tubes rotate along with the rotary drum. The heat of drying is provided mainly by the condensation heat of the compressed steam in the heating tubes. The wet material is fed continuously at a uniform rate and is tumbled and agitated by the cylinder's rotation. A steam tube rotary dryer has a slope rotating cylinder to move the material being dried from the charging end to the discharging end by gravity. Finally, the dried material is discharged from the outlet. During the travel inside the cylinder, the material is heated by contact with radiant heat from the steam tubes' surfaces. The heat exchange mode inside the dryer is counter-current to achieve a minimum exergy loss following heat exchange. This heat exchange mode becomes one of the benefits of steam tube rotary dryers compared with fluidized bed dryers, usually involving a cocurrent operation.

Further, E3.3 shows the schematic diagram of the proposed algae drying based on heat circulation technology with steam recirculation. The steam exhausted from the main evaporator (HX3) is split into purged and recirculated steam (S12). The amount of the purged steam is equivalent to the water evaporated from the wet brown algae. By adopting steam recirculation, it is considered that the heat exchange inside the evaporator can be more uniform, and the drying rate can be faster. The total energy input in the drying system consists of the compressor and blower work.

Continued

Example of algae drying—cont'd

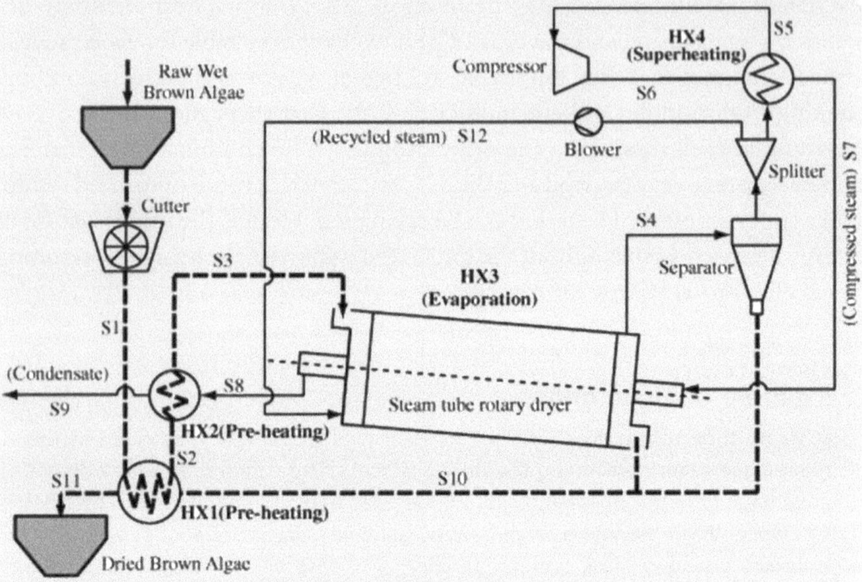

Figure E3.3 Schematic diagram of the innovative drying process based on heat circulation technology.

In the proposed drying processes employing HC, drying is divided into three subsequent stages: preheating, evaporation, and superheating. The wet brown algae enter the cutting treatment before entering the preheating phase to achieve a uniform size and improve their dynamics in the dryer; hence, a better and more uniform heat transfer could be performed. The wet alga is heated to a certain temperature by the compressed vapor flowing out from the steam tube rotary dryer in the preheating stage. In this stage, sensible heat exchange is conducted between wet algae as the cold stream and compressed vapor as the hot stream.

Subsequently, the wet alga enters the main evaporation stage conducted in the steam tube rotary dryer. The wet algae enter the dryer from the inlet at the higher end of the dryer, while the compressed vapor flows from the other side, which is the discharge end of the dryer. In the steam tube rotary dryer, the compressed water's condensation heat is exchanged with the algae moisture's evaporation heat. The drying alga flows down through the dryer and is finally exhausted through the outlet at the lower end of the dryer. The evaporated water will be exhausted and delivered to a filter before its exergy rate is raised by compression; therefore, the stream's physical properties will change,

Example of algae drying—cont'd

including the saturation temperature, resulting in the possibility of latent heat recovery. The compressed evaporated water acts as a hot stream, providing the heat source in all of the drying stages, and hence, no additional heater is required.

The superheating stage is introduced in the proposed system to preserve a complete vapor phase (vapor quality equals 1.0), avoiding any condensation during compression inside the compressor. It is well known that condensation inside the compressor could lead to a breakage or a shorter compressor's working life.

The new drying process approach can significantly reduce the energy required compared with the conventional drying process based on heat cascade-utilization technology. Exergy recovery through exergy elevation and heat pairing applied in heat circulation technology can recuperate the energy involved in drying effectively. As the drying energy could be reduced significantly, it also means that algae drying application could increase the fossil energy ratio and the energy density of algae. Hence, algae's economic performance can be improved significantly, leading to the possibility of broader and larger utilization of algae for feedstocks and as an energy source. Furthermore, the proposed drying processes can also decrease the CO_2 emitted during drying. The steam recirculation in the brown algae drying process does not significantly impact total energy consumption due to small blower work circulating the steam. To further reduce the energy consumption in algae utilization, including pretreatment, conversion, utilization, and posttreatment processes, a combination of the heat circulation technology and process integration technologies is considered very important as the next task to be evaluated. This is owing to a further minimization of total exergy loss in the whole process.

3.7 Integrated cogeneration system from biomass adopting enhanced process integration: an example

Fig. 3.23 denotes a schematic diagram of the integrated cogeneration system from biomass algae published earlier [55]. Enhanced process integration technology discussed above is employed to optimize heat recovery in the

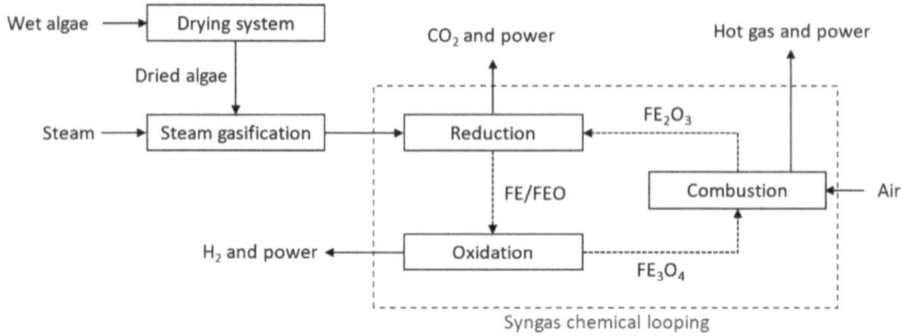

Figure 3.23 Schematic diagram of the cogeneration system employing heat recovery and material recovery system.

system, thus increasing the power generation system's efficiency. The integrated system consists of drying, steam gasification, chemical looping, and power generation. Firstly, fresh wet algae with a moisture content of 85% were cut to a more uniform size to improve the heat transfer during the drying process. The mass flow of wet algae was assumed to be 100 th^{-1}.

Wet brown macroalgae were first subjected to a drying process to improve their characteristics. The dried algae were then fed into the gasification process, in which they were reacted with steam to produce syngas. A dual circulating fluidized-bed gasifier consisting of a gasifier and a combustor operated at near-atmospheric pressure was used to convert algae. This technology has a higher carbon conversion efficiency, higher conversion rate, high mass and heat transfer, and high efficiency. The chemical looping process used in this system utilized three reactors: a fuel reactor (reduction), a steam reactor (oxidation), and an air reactor (combustion). Such a configuration is commonly referred to as syngas chemical looping (SCL). The syngas produced in the gasifier was reacted with OCs (iron oxide), which had CO_2 and water stream as emissions. In the oxidation stage, the reduced particles react with steam to produce an H_2-rich gas stream, flowing out together with the remaining steam. The OCs were then recovered to their initial state by combustion with air. The heat generated in each reactor was recovered by using expanders to produce power. The total energy efficiency (η_{total}) is calculated as the sum of the power generation efficiency (η_{power}) and the H_2 production efficiency (η_{H2}), as follows:

$$\eta_{power} = \frac{W_{net}}{m_{dried\ algae}\ x\ LHV_{algae}} \tag{3.34}$$

$$\eta_{H2} = \frac{m_{H2} x\ LHV_{H2}}{m_{dried\ algae}\ x\ LHV_{algae}} \tag{3.35}$$

$$\eta_{total} = \eta_{power} + \eta_{H2} \tag{3.36}$$

References

[1] Smith JM. Introduction to chemical engineering thermodynamics. 1950. https://doi.org/10.1021/ed027p584.3. n.d.
[2] Sussman MV. Elementary general thermodynamics. Addison-Wesley Publishing Company; 1972.
[3] Moran MJ. Engineering thermodynamics: fundamentals. 2017. https://doi.org/10.4324/9781315119717.
[4] Denbigh KG. The principles of chemical equilibrium. Cambridge University Press; 1981. https://doi.org/10.1017/cbo9781139167604.
[5] Ptasinski KJ. Efficiency of biomass energy - an exergy approach to biofuels. Power Biorefineries 2017;89. https://doi.org/10.1002/cite.201770067.
[6] Singh K, Tollner EW, Mani S, Risse LM, Das KC, Worley J. Transforming solid wastes into high quality bioenergy products: entropy analysis. In: 2008 Proc. 16th Annu. North Am. Waste to Energy Conf. NAWTEC16, ASME; 2008. p. 131−40. https://doi.org/10.1115/nawtec16-1924.
[7] Haseli Y. Exergy. Entropy Anal. Therm. Eng. Syst. Elsevier; 2020. p. 169−83. https://doi.org/10.1016/B978-0-12-819168-2.00011-8.
[8] Babu V. Fundamentals of engineering thermodynamics. 2019. https://doi.org/10.1201/9780367816087.
[9] Szargut J. International progress in second law analysis. Energy 1980;5:709−18. https://doi.org/10.1016/0360-5442(80)90090-0.
[10] Amelio A, Van De Voorde T, Creemers C, Degrève J, Darvishmanesh S, Luis P, et al. Comparison between exergy and energy analysis for biodiesel production. Energy 2016;98:135−45. https://doi.org/10.1016/j.energy.2016.01.018.
[11] Gundersen T. The concept of exergy and energy quality. 2009.
[12] Aspelund A, Berstad DO, Gundersen T. An extended pinch analysis and design procedure utilizing pressure based exergy for subambient cooling. Appl Therm Eng 2007;27:2633−49. https://doi.org/10.1016/j.applthermaleng.2007.04.017.
[13] Bejan A. Exergy analysis of thermal, chemical and metallurgical processes. Int J Heat Fluid Flow 1989;10:87−8. https://doi.org/10.1016/0142-727x(89)90062-3.
[14] Szargut J. Exergy method: technical and ecological applications, vol. 18; 2005.
[15] Szargut J, Styrylska T. Approximate evaluation of the exergy of fuels. Brennst Waerme Kraft 1964;16:589−96.
[16] Song G, Shen L, Xiao J. Estimating specific chemical exergy of biomass from basic analysis data. Ind Eng Chem Res 2011;50:9758−66. https://doi.org/10.1021/ie200534n.
[17] Gourmelon S, Thery-Hetreux R, Floquet P, Baudouin O, Baudet P, Campagnolo L. Exergy analysis in ProSimPlus® simulation software: a focus on exergy efficiency evaluation. Comput Chem Eng 2015;79:91−112. https://doi.org/10.1016/j.compchemeng.2015.02.014.
[18] Szargut J, Morris D, Steward F. Exergy analysis of thermal, chemical, and metallurgical processes. Hemisphere Publishing Corporation; 1988.

[19] Cornelissen RL. Thermodynamics and sustainable development the use of exergy analysis and the reduction of irreversibility. 1997.
[20] Labidi J, Boulet E, Paris J. On intrinsic exergy efficiency and heat pumps. Chem Eng Res Des 2000;78:180–3. https://doi.org/10.1205/026387600527202.
[21] Cornelissen RL, Hirs GG. Exergy analysis of cryogenic air separation. Energy Convers Manag 1998;39:1821–6. https://doi.org/10.1016/s0196-8904(98)00062-4.
[22] Kotas TJ. The exergy method of thermal plant analysis. Elsevier; 1985. https://doi.org/10.1016/c2013-0-00894-8.
[23] Horlock JH. The rational efficiency of power plants and their components. J Eng Gas Turbines Power 1992;114:603–11. https://doi.org/10.1115/1.2906633.
[24] Tsatsaronis G. Thermoeconomic analysis and optimization of energy systems. Prog Energy Combust Sci 1993;19:227–57. https://doi.org/10.1016/0360-1285(93)90016-8.
[25] Lazzaretto A, Tsatsaronis G. SPECO: a systematic and general methodology for calculating efficiencies and costs in thermal systems. Energy 2006;31:1257–89. https://doi.org/10.1016/j.energy.2005.03.011.
[26] Rosen MA, Bulucea CA. Using exergy to understand and improve the efficiency of electrical power technologies. Entropy 2009;11:820–35. https://doi.org/10.3390/e11040820.
[27] Ahmadi P, Dincer I. Energy optimization. Compr. Energy Syst., vol. 1–5. Elsevier Inc.; 2018. p. 1085–143. https://doi.org/10.1016/B978-0-12-809597-3.00135-8.
[28] Demirel Y, Gerbaud V. Thermoeconomics. Nonequilibrium thermodyn. Elsevier; 2019. p. 267–94. https://doi.org/10.1016/b978-0-444-64112-0.00005-8.
[29] Kanoglu M, Dincer I, Cengel Y. Exergy for better environment and sustainability. Environ Dev Sustain A Multidiscip Approach Theory Pract Sustain Dev 2009;11:971–88.
[30] Saidur R, Boroumandjazi G, Mekhilef S, Mohammed HA. A review on exergy analysis of biomass based fuels. Renew Sustain Energy Rev 2012;16:1217–22. https://doi.org/10.1016/j.rser.2011.07.076.
[31] Rao MS, Singh SP, Sodha MS, Dubey AK, Shyam M. Stoichiometric, mass, energy and exergy balance analysis of countercurrent fixed-bed gasification of post-consumer residues. Biomass Bioenergy 2004;27:155–71. https://doi.org/10.1016/j.biombioe.2003.11.003.
[32] Chen ZS, Wang LQ. Energy and exergy analysis of gas production from biomass intermittent gasification. J Renew Sustain Energy 2013;5:063141. https://doi.org/10.1063/1.4857395.
[33] Buentello-Montoya D, Zhang X. An energy and exergy analysis of biomass gasification integrated with a char-catalytic tar reforming system. Energy and Fuels 2019;33:8746–57. https://doi.org/10.1021/acs.energyfuels.9b01808.
[34] Costa VAF, Tarelho LAC, Sobrinho A. Mass, energy and exergy analysis of a biomass boiler: a Portuguese representative case of the pulp and paper industry. Appl Therm Eng 2019;152:350–61. https://doi.org/10.1016/j.applthermaleng.2019.01.033.
[35] Peters JF, Petrakopoulou F, Dufour J. Exergetic analysis of a fast pyrolysis process for bio-oil production. Fuel Process Technol 2014;119:245–55. https://doi.org/10.1016/j.fuproc.2013.11.007.
[36] Prasakti L, Pratama SH, Fauzi A, Pradana YS, Rochmadi BA. Exergy analysis of conventional and hydrothermal liquefaction-esterification processes of microalgae for biodiesel production. Open Chem 2020;18:874–81. https://doi.org/10.1515/chem-2020-0132.
[37] Puig-Arnavat M, Bruno JC, Coronas A. Review and analysis of biomass gasification models. Renew Sustain Energy Rev 2010;14:2841–51. https://doi.org/10.1016/j.rser.2010.07.030.

[38] Mutlu ÖÇ, Zeng T. Challenges and opportunities of modeling biomass gasification in aspen plus: a review. Chem Eng Technol 2020;43:1674–89. https://doi.org/10.1002/ceat.202000068.
[39] Dahlquist E, Mirmoshtaghi G, Larsson EK, Thorin E, Yan J, Engvall K, et al. Modelling and simulation of biomass conversion processes. 2013 [n.d].
[40] Klemes J, Friedler F, Bulatov I, Varbanov P. Sustainability in the process industry: integration and optimization: integration and optimization. McGraw-Hill Education; 2010.
[41] Foo DCY, Ng DKS. Process integration for cleaner process design. Handb. Process Integr. Minimisation energy water use, waste Emiss. Elsevier Inc.; 2013. p. 443–60. https://doi.org/10.1533/9780857097255.4.443.
[42] Leites IL, Sama DA, Lior N. The theory and practice of energy saving in the chemical industry: some methods for reducing thermodynamic irreversibility in chemical technology processes. Energy 2003;28:55–97. https://doi.org/10.1016/S0360-5442(02)00107-X.
[43] Kenney WF. Energy conservation in the process Industries. Elsevier Inc.; 2012. https://doi.org/10.1016/C2009-0-21865-4.
[44] Kemp I. Pinch analysis and process integration. Elsevier Ltd; 2006. https://doi.org/10.1016/B978-0-7506-8260-2.X5001-9.
[45] Higa M, Freitas AJ, Bannwart AC, Zemp RJ. Thermal integration of multiple effect evaporator in sugar plant. Appl Therm Eng 2009;29:515–22. https://doi.org/10.1016/j.applthermaleng.2008.03.009.
[46] Linnhoff B, Hindmarsh E. The pinch design method for heat exchanger networks. Chem Eng Sci 1983;38:745–63. https://doi.org/10.1016/0009-2509(83)80185-7.
[47] Fujimoto S, Yanagida T, Nakaiwa M, Tatsumi H, Minowa T. Pinch analysis for bioethanol production process from lignocellulosic biomass. Appl Therm Eng 2011;31:3332–6. https://doi.org/10.1016/j.applthermaleng.2011.06.013.
[48] Li Z, Ma L, Jia X, Ren J. Pinch analysis for sustainable process design and integration. Towar. Sustain. Chem. Process. Elsevier; 2020. p. 275–91. https://doi.org/10.1016/B978-0-12-818376-2.00010-7.
[49] Gundersen T. Heat integration: targets and heat exchanger network design. Handb. Process Integr. Minimisation energy water use, waste Emiss. Elsevier Inc.; 2013. p. 129–67. https://doi.org/10.1533/9780857097255.2.129.
[50] Bagajewicz M. Pinch and minimum utility usage. [n.d].
[51] Pinch analysis: for the efficient use. [n.d].
[52] Rokni M. Introduction to pinch technology division of energy section introduction to pinch technology. Technical University of Denmark; 2016.
[53] El-Halwagi MM, Yee Foo DC. Process synthesis and integration. Kirk-othmer Encycl. Chem. Technol. Hoboken, NJ, USA: John Wiley & Sons, Inc.; 2014. p. 1–24. https://doi.org/10.1002/0471238961.1618150308011212.a01.pub2.
[54] Aziz M, Oda T, Kashiwagi T. Enhanced high energy efficient steam drying of algae. Appl Energy 2013;109:163–70. https://doi.org/10.1016/j.apenergy.2013.04.004.
[55] Zaini IN, Nurdiawati A, Aziz M. Cogeneration of power and H_2 by steam gasification and syngas chemical looping of macroalgae. Appl Energy 2017. https://doi.org/10.1016/j.apenergy.2017.06.071.

CHAPTER 4

Proposed integrated system from black liquor

Arif Darmawan[1], Muhammad Aziz[2] and Koji Tokimatsu[3]
[1]Agency for the Assessment and Application of Technology (BPPT), Puspiptek Serpong, Tangerang Selatan, Indonesia; [2]Institute of Industrial Science, The University of Tokyo, Meguro-ku, Tokyo, Japan; [3]Department of Transdisciplinary Science and Engineering, Tokyo Institute of Technology, Midori-ku, Yokohama, Kanagawa, Japan

Enormous energy potential as a feedstock for biofuels can be obtained from residues of a biomass-based pulp mill. During the industrial-scale manufacturing process, wood chips are cooked at high temperatures and white liquor is used to dissolve lignin. Waste cooking liquor, called black liquor (BL), contains inorganic cooking chemicals and combustible materials [1]. BL is a viscous alkaline solution and has various compositions depending on wood properties and cooking conditions. Generally, it contains lignin as the main composition and organic compounds of cellulose, hemicellulose, and alkaline salts such as sodium hydroxide and sodium sulfide [2]. In the pulp and paper industry, about 170 million tonnes of BL are processed every year (in terms of dry matter), and a single pulp can produce 1.7−1.8 t-BL/t-pulp (dry matter) with a potential energy source of 250−500 MW/mill. The recovery of energy from the BL has the potential to generate both the electricity and steam required for the continuous pulp mill operation.

Based on data, a pulp mill consumes electricity about 700 kWh per ton of pulp [3]. Ideally, in the modern pulp mill industry, a factory system can independently provide internal energy demands for daily operation. The energy sources can come from the conventional combustion of BL in a recovery boiler (Fig. 4.1) after the drying or evaporation process removing the BL water content. Without proper drying, the combustion efficiency decreases, leading to an unsustainable recovery process. BL produced from an extraction plant has a moisture content ranging from 83% to 87% on a wet basis (wb) and must be concentrated to a certain level for combustion [3]. Although this way is a proven method, it has a major disadvantage due to low energy efficiency [4].

Due to excessive water content, evaporation or drying of BL requires a lot of energy. It becomes the pulp mill factory's biggest energy consumer by using 50% of the entire internal steam demand [5]. In traditional systems,

Innovative Energy Conversion from Biomass Waste
ISBN 978-0-323-85477-1
https://doi.org/10.1016/B978-0-323-85477-1.00005-1

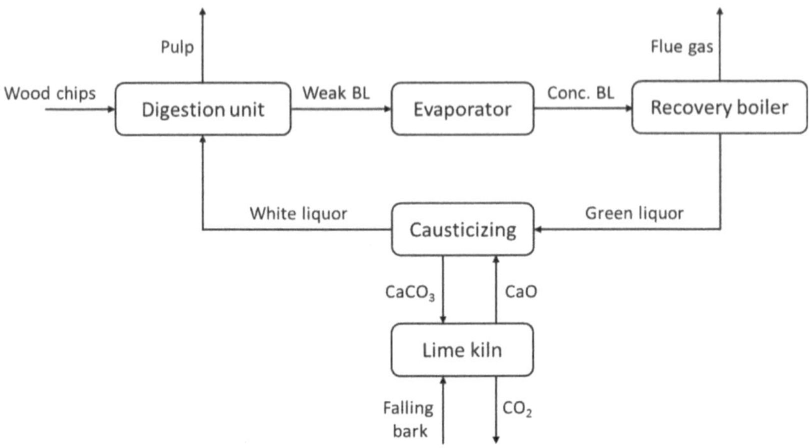

Figure 4.1 Energy recovery cycle in Kraft pulping process.

BL drying adopts multiple-effect evaporator (MEE). Since MEE systems can concentrate BL to only about 50% solids, a high-solid concentrator is then employed to achieve a solid concentration higher than 65% or 70%. The energy allocation required in MEE and high-solid concentration is approximately 60% and 40%, respectively. The MEE can be either forward-feed arrangement or backward-feed arrangement. The multiple-effect evaporator operated in the backward-feed arrangement is shown in Fig. 4.2, in which steam and BL flow in the opposite direction. This arrangement is beneficial and gives higher capacity than the forward-feed when the concentrated liquid is viscous since the viscous flow is at a higher temperature being in the first effect. For the conventional boiler combustion, BL evaporations are performed until the solid concentration increases to approximately 75–85 wt% wb [6]. In order to improve the total energy efficiency of pulp milling energy, efficient BL drying is urgently required.

4.1 Conventional energy recovery from black liquor

Since BL is available within the pulp mill, it is advantageous and can be conveniently processed into biomass fuel without additional transportation costs. It is the main biomass resource in some countries with large pulp and paper industries. Today, this large amount of BL's energy potential is utilized to meet the pulp mills' energy demands. Nevertheless, a portion of the total electricity demand is met by importing electricity. Generally, the

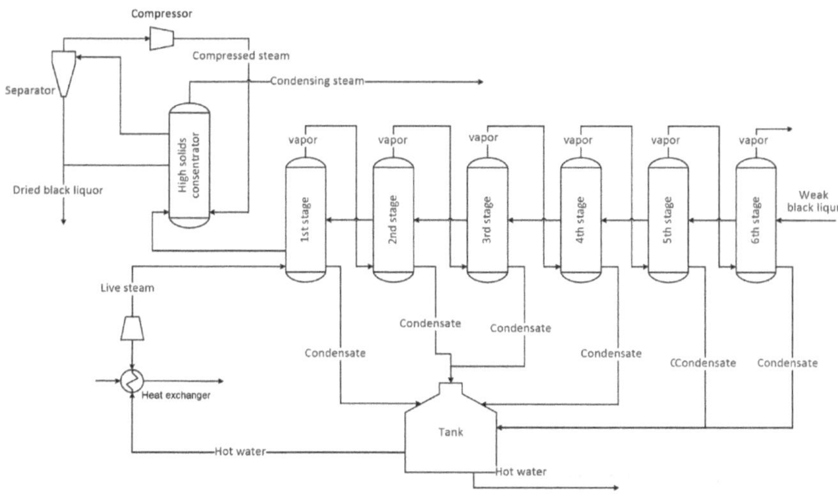

Figure 4.2 Backward-feed multiple-effect evaporator combined with a high-solid concentrator.

chemical recovery process for BL consists of the following four main steps: BL concentration, BL combustion (recovery boiler), recausticizing, and calcification (lime burning) [3]. For biodiesel production, BL recovery is divided into three processes: lignin removal, gasification, and tall oil processing [7].

This section describes the technologies available for the production of electricity and H_2 from BL and discusses these bio-based recovery performances from the energy perspective. The first conventional system shown in Fig. 4.3 uses a recovery boiler primarily to produce steam. This method is commonly adopted in modern pulp mills. First, a weak BL with high water content (about 85 wt% wet basis (wb)) is passed through a MEE to increase the solid content (up to 70–75 wt% wb) and make it suitable for combustion. Most factories use MEE by adopting indirect heat from steam.

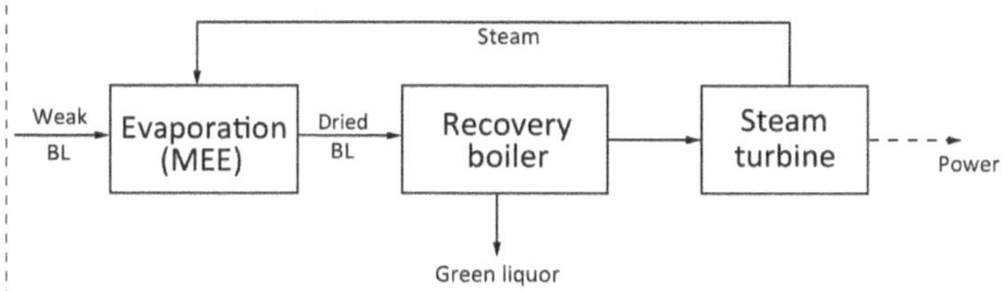

Figure 4.3 Integrated system of conventional recovery boiler in most pulp mill industries.

Another approach is direct contact evaporation (DCE) that uses the exhaust gas from the recovery boiler to concentrate the liquid BL.

Since BL evaporation is the largest steam consumer in a pulp mill factory, MEE is the best choice to maximize efficiency. MEE systems can concentrate BL to about 50% solids. A DCE is employed to achieve a final concentration of 70%–80% using waste gases from the recovery boiler [3]. MEE systems that can produce higher solids content also become major consent to eliminate the DCE use for optimum performance. During the combustion phase in the recovery boiler, heat is obtained and high-temperature steam is generated to produce electricity. Basically, a higher solids content is preferable in a recovery boiler, leading to overall better performance. Superheaters and economists are generally adopted to increase the efficiency of a boiler system. The remaining steam is then sent to the pulp mill system for supporting other processes. During BL solid combustion, the organic components are burned to produce heat, and the inorganic chemicals in the process are dissolved. The solvent is removed from the bottom of the boiler for further refinement in the recaustification stage.

Although the BL gasification process remains under development, the BL gasification combined cycle (BLGCC) schemes have been studied and analyzed to obtain the best bio-based system for BL energy recovery. BL gasification is a promising technology and maybe a future alternative or complement to traditional recovery boilers in the pulp mill industry. This process produces synthetic gases (gas mixtures containing H_2, CO, etc.) that can be used as fuel for internal combustion engines or generate steam for electricity. Fig. 4.4A and B show BLGCC without and with CO_2 capture, respectively. These BLGCC systems are based on data from business-as-usual patterns in an integrated pulp and paper mill [8,9]. The system mainly includes gasifiers, gas purification units, gas turbines, heat recovery steam generators, steam turbines, and condensers. Dry BL is initially gasified, and then the syngas is combusted in a gas turbine to generate power. A heat recovery steam generator (HRSG) is used in a combined cycle scheme similar to an integrated gasification combined cycle (IGCC). In BLGCC with CO_2 capture/sequestration, a CO_2 recovery unit is installed as a precombustion technique to remove CO_2 before the combustion. Elzimar et al. [8] reported, due to the energy consumption of CO_2 capture components, the net thermal efficiency was calculated to be 6.57% lower than BLGCC without CO_2 capture. However, compared to the recovery of BL by combustion, the configuration of BLGCC can significantly increase the

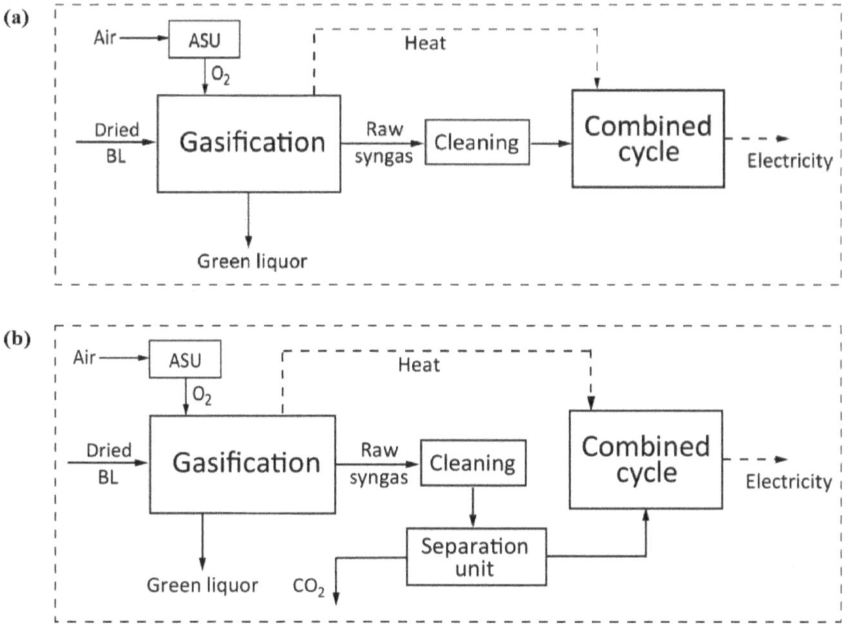

Figure 4.4 Integrated system of (A) black liquor gasification combined cycle (BLGCC) and (B) BLGCC with CO_2 capture.

efficiency of the system (>10%), and the pulp mill can shift from the power importer to the power exporter [4].

Although the production of H_2 from biomass has been widely investigated, the production of H_2 from BL is rarely studied. Anderson and Harvey [10] reported the possibility of H_2 being produced in a pulp mill by gasifying BL feedstock. The system mainly consists of gasification, gas purification, and gas separation, as shown in Fig. 4.5. The system is able to capture and store significant CO_2 emissions up to 100%. After gasification and heat recovery processes, H_2S is absorbed into MDEA (methyl

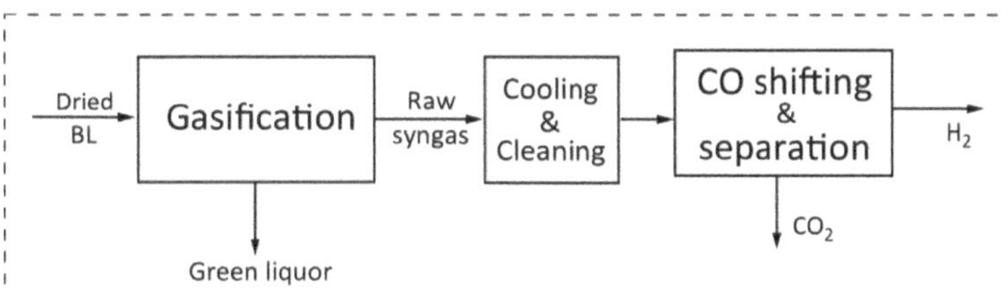

Figure 4.5 Scheme of an H_2 production unit with CO_2 separation from gasified black liquor (BL).

Table 4.1 Performance of different integrated systems.

	Carbon capture (%)	H_2 (%LHV)	Net power (%LHV)	(Efficiency (%LHV)
MEE + conventional recovery boiler	–	–	9–14	9–14
BLGCC	–	–	31	31
MEE + BLGCC	–	–	24	24
BLGCC with CO_2 capture	90	–	24.3–28	24.3–28
MEE + BLGCC with CO_2 capture	90	–	22.4	22.4
MEE + conventional H_2 production	90	37.5	–	37.5

diethanolamine), reabsorbed into white liquor, and reused in the pulping process. The syngas is shifted in two water-gas shift reactors to achieve maximum conversion of CO into H_2.

For a brief performance evaluation, Table 4.1 compares lower heating value (LHV)-based efficiency between several systems that use BL for power or H_2 production. The conventional BL recovery system corresponding to Fig. 4.3 uses a conventional recovery boiler called the Tomlinson Boiler. The BL is dried in MEE and then burned in the boiler. The system generates electricity using the back-pressure/extraction steam turbine conventional cycle, leading to relatively low energy efficiency. In contrast, the BLGCC showed in Fig. 4.4A offers 31% higher energy efficiency, which is twice as much as conventional systems. Also, the combustion of the syngas in gas turbines generates more power. In this case, the gas turbine exhaust is used to produce high-pressure steam for the back-pressure steam turbine cogeneration system. Integrating MEE into the investigated system can reduce efficiency to 24%, which requires energy for the evaporation process. In addition, BLGCC with CO_2 capture would require a high amount of energy to accommodate the separation reactions. If the same system includes MEE and BL drying, the overall efficiency will be 24.3%–28% without CO_2 capture and 22.4% with CO_2 capture. In general, MEE consumes larger amounts of energy as auxiliary power and a larger amount of heat to generate steam. Finally, the conventional production of H_2 uses a water-gas shift reaction to produce CO_2 and H_2, which also requires energy as well. In contrast, the system could relatively achieve higher total efficiency (37.5%) than other processes.

4.2 Bio-based proposed system employing evaporation, gasification, and combined cycle

Here an improved process integration technology is adopted to reduce exergy loss throughout the integrated system greatly. It is a combination of two technologies: exergy recovery and process integration, as discussed in the previous chapter. Exergy recovery refers to the circulation of energy involved in a given process by combining an increase in exergy rate and efficient thermal matching. However, for some reasons, such as physical and chemical changes and heat exchanger limitations, not all of the heat involved in the process can be recovered internally. Therefore, to reduce the exergy loss, process integration is used to facilitate the use of heat and minimize the total exergy loss in the integrated system. Fig. 4.6 shows the first conceptual diagram of the general integrated system proposed for the use of BL. The study focuses on two main points: the evaporation system for BL and the integrated system for producing electricity from BL, including evaporation, gasification, and combined cycle.

We cannot show further details about exergy recovery and process integration in the following figure, so detailed information about each module's issues is provided in the next section. The proposed integrated system mainly consists of BL evaporation, BL gasification, and combined cycle. Solid, dashed, and dotted lines represent material, heat, and electricity streams, respectively. As the main BL evaporator, a rotary steam tube evaporator with an internal heat exchanger is selected. Its advantages include a large heat transfer area, excellent thermal efficiency, high handling capacity, and continuous operation ease.

The high concentrated BL from the evaporation system is then fed to the gasification module. The solid BL is converted into syngas at high temperature, which is made up of fuel gases, including H_2 and CO. The

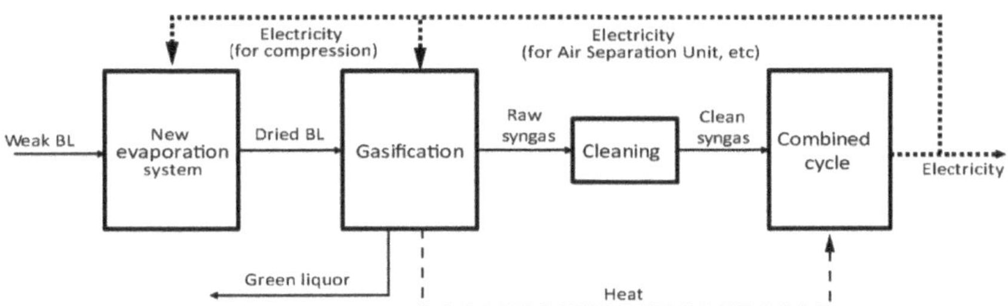

Figure 4.6 Conceptual diagram of the proposed black liquor integrated system for power generation.

produced gas is used as a combustion fuel to generate electricity. The complex reactions involving pyrolysis, volatilization, combustion, and char gasification occur continuously in the gasifier and the syngas generated. The hot syngas then flows to a heat recovery module and supports the combined cycle process (preheating purpose). Furthermore, particulate matter and sulfur are removed from the synthesis gas. After purification, the syngas enters the combustion chamber, producing high-temperature and high-pressure gas for the gas turbine. Also, because the temperature of the gas turbine's flue gas is relatively high, the heat is recovered to generate steam in the heat recovery steam generator (HRSG) and utilized to run steam turbines to generate power.

4.2.1 Proposed black liquor evaporation system

As mentioned previously, most pulp factories use MEEs to increase BL concentrations. Before sending it to the boiler, the water content of the BL must be removed to maximize the energy recovery efficiency during combustion. MEE has the highest steam consumption among other modules of the pulp mill. Depending on the type of MEE, steam can be used for direct and indirect contact. Fig. 4.7 denotes the schematic process flow diagram of the proposed BL evaporation. The proposed system does not use MEE but adopts the principle of self-heat exergy recovery. The BL evaporation system is designed to consist of three consecutive steps: BL preheating, BL evaporation, and steam superheating. In the first stage, the liquid BL is preheated by condensed and compressed steam heat, which flows from the steam tube rotary evaporator (HX2). Sensible heat is exchanged between the compressed vapor and the liquid BL.

Subsequently, the weak or liquid BL flows to the second stage, main evaporation, which occurs in the rotary steam tube evaporator. A weak BL enters the inlet evaporator, while the compressed vapor (F6) flows from the opposite side, thus facilitating counter-current heat exchange between the two streams. In this process (HX2), the evaporation's heat is provided by the compressed water's condensation. The dried BL (F9) flows out from the evaporator outlet for the next conversion process. The evaporator's exhaust vapor is split into recirculated vapor (F10) and purged/compressed vapor (F4). The latter is compressed to increase the exergy rate before its heat is exchanged in the evaporator (HX2). Raising the exergy level means improving the quality of energy that can be used for the next process. The steam physical properties transform by compression, including its saturation point, leading to latent heat recovery.

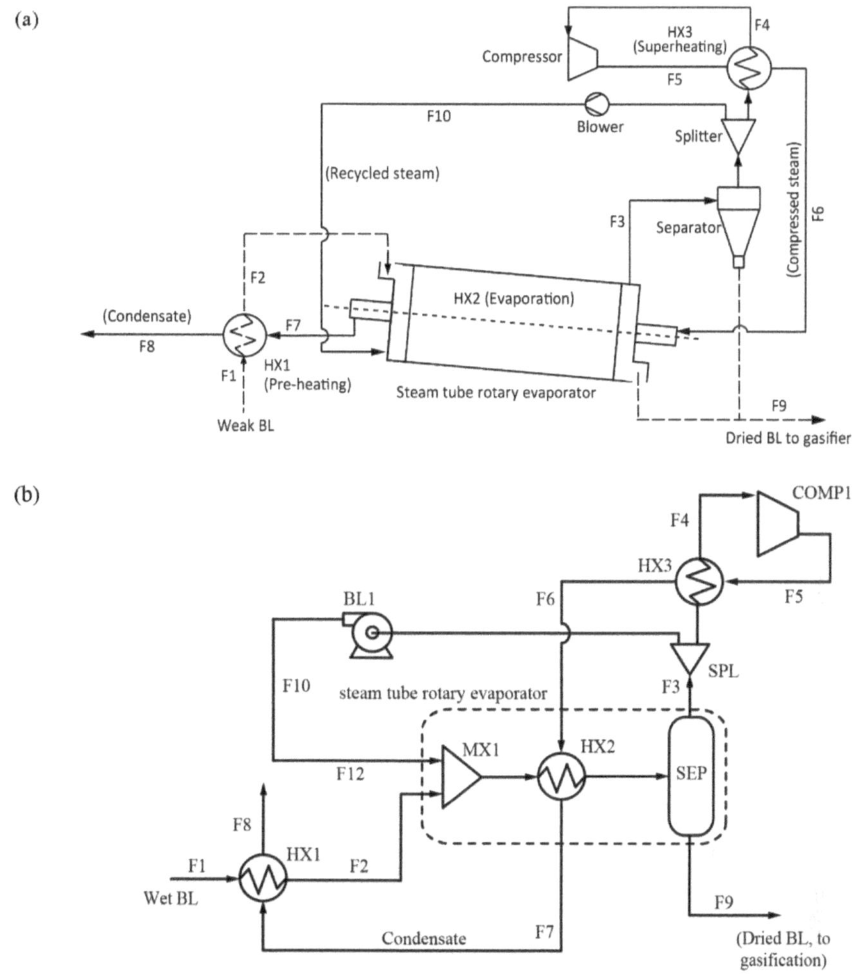

Figure 4.7 (A) Arrangement and (B) process flow diagram of proposed black liquor (BL) evaporation scheme.

On the other hand, the recycle vapor (F10) is recirculated to the evaporator to improve the performance of heat transfer by evaporation and the BL dynamics. In this way, faster and more uniform evaporation can be obtained. With these goals in mind, the recycled steam's mass flow rate must be greater than the flow rate of the BL. The final stage, superheating, is used in the evaporation system to ensure a perfect gas phase during compression. The main purpose is to avoid any steam condensation. Condensation inside the compressor can reduce or shorten the life of the compressor. BL also contains volatile substances, some of which evaporate after the water evaporates. However, the amount of material that evaporates depends largely on the pulp's origin and the evaporation conditions (pressure and temperature). Therefore, superheating is very important to avoid the condensation of steam and the condensation of volatile substances. In practice, the superheating conditions should be investigated and evaluated to enhance the overall drying performance.

4.2.2 Integrated process for gasification and power generation

Fig. 4.8 illustrates the process flow diagram of the proposed integrated system for power generation, covering gasification and combined cycle processes. In the system, new process integration is proposed to reduce losses. The gasifier (GF) uses oxygen (F28) as an oxidant for gasification reactions. Therefore, the air separation unit (ASU) is installed. If the generated oxygen (F28) is used in gasification, the generated nitrogen (F29) is injected into the gas turbine (GT) combustor. Nitrogen supplied reduces NOx formation by lowering peak flame temperature and excess water and oxygen concentrations produced during combustion. The generated syngas (F10) is then cooled in a heat exchanger, where the heat is recovered to preheat the steam turbine. This heat recovery phase is an important part of the calculation to avoid thermal exergy losses in the overall integrated

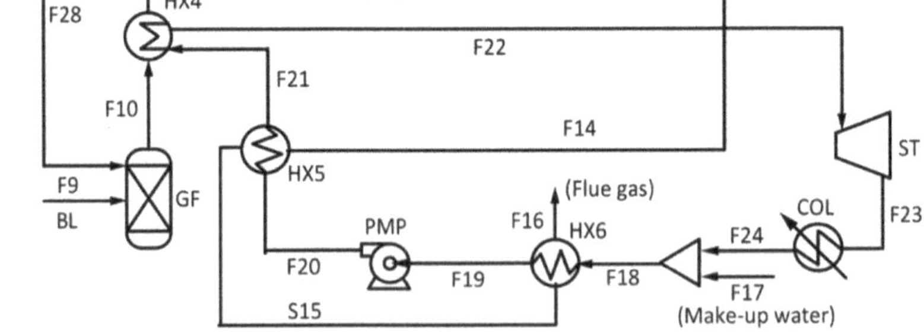

Figure 4.8 Proposed process flow diagram covering black liquor (BL) gasification and combined cycle.

system. The H_2O and organic matter such as tar contained in the syngas flow can condense at low temperatures; therefore, the temperature is kept above 300°C before the gas purification stage. The cleaned syngas go to the combustor to produce high-temperature and high-pressure gas and then expand in the GT, generating the mechanical energy necessary to rotate the generator. Subsequently, the superheated steam (F21) flows from the superheated steam (F21) to the superheater (HX4) and the steam turbine (ST) to generate additional power.

Table 4.2 shows the calculated conditions for gasification and combined cycle processes. An oxygen-blown pressurized entrained flow gasifier featuring a centrally mounted burner with a coannular nozzle at the top of the vessel is used. The nozzle is used to inject small amounts of BL, oxygen, and nitrogen into a ceramic-lined reactor with a pressure of 2.7–2.9 MPa BL initially atomized by oxygen at the nozzle outlet; subsequently, gasification occurs, producing smelt and hot syngas.

Table 4.2 Assumed conditions during gasification and combined cycle.

Parameters	Value
Gasification [11]	
Temperature (°C)	1036
Gasifier pressure (MPa)	2.97
Gasifier type	Entrained flow gasifier
Produced gas (after gas cleaning)	H_2 (35.2%), CO (25.7%) CO_2 (35.7%), and minor gas components (CH_4, H_2S, N_2)
Combustor and gas turbine	
Discharge pressure (MPa)	0.15
Combustor pressure drop (%)	2
Gas turbine isentropic efficiency (%)	90
HRSG and steam turbine	
HRSG outlet pressure (MPa)	20
HRSG pressure drop (%)	1
Steam turbine isentropic efficiency (%)	90
Minimum vapor quality at ST outlet (%)	90

The net power efficiency of the integrated system is calculated as follows:

$$\eta_{net} = \frac{P_{gross} - P_{int}}{\dot{m}_{BL} \times LHV_{BL}}, \tag{4.1}$$

where P_{gross}, P_{int}, $\dot{m}_{BL}$, and LHV_{BL} are gross generated power, internally consumed power, the mass flow rate of BL, and LHV of BL, respectively.

4.2.3 General conditions during simulation

In order to design and perform calculations related to the proposed integrated system's mass and energy balance, we used the Aspen Plus V8.8 Process Simulator (Aspen Technology, Inc.). The weak BL flow rate that entered the process modules is 348.12 t/h (considering average pulp production of 730 t/d) (See the composition in Table 4.3). Some additional assumptions were made: (i) The evaporator system consists of a mixer, heat exchanger, and separator; (ii) The minimum temperature difference in all heat exchangers is 30 K; (iii) An entrained flow gasifier consists of a mixer and a conversion reactor; (iv) Ambient temperature is 25°C; (v) Adiabatic efficiency of both compressor and blower is 90%; (vi) There is no heat loss; (vii) There is no air contamination inside the evaporator; and (viii) In the evaporator, heat is transferred from a hot material to BL particles through a heat transfer tube. Also, used air contains 79 mol% of nitrogen and 21 mol% of oxygen.

4.2.4 Evaporation system performance

Fig. 4.9 shows the BL solid concentration's effect on the total work required and the compressor outlet pressure during evaporation. In general, the higher BL concentration, the better the mass balance and thermal bond, which reduces the total work required during evaporation. The BL concentration requires the total work to reach a concentric BL of 60 wt% wb higher than that required to reach a solid concentration above 60 wt% wb.

Table 4.3 Composition of black liquor (BL) used herein.

Properties	Value
Total solids (wt% wb)	15
Water content (wt% wb)	85
Components (wt% db) [11]	C: 27.50; H: 3.75; O: 39.35; N: 0.07; Cl: 0.16; Na: 19.85; K: 3.12, S: 6.20
LHV (MJ/kg, db)	12.13

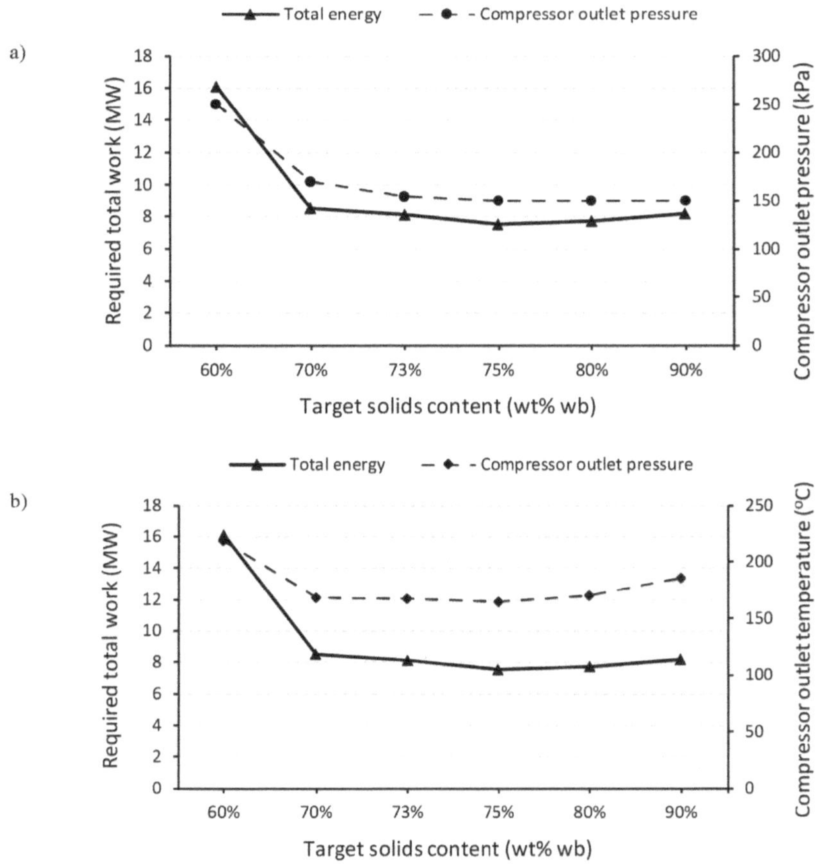

Figure 4.9 Effect of target solid content on evaporation performance: (A) required work and compressor outlet pressure, (B) required work and compressor outlet temperature.

This is mainly due to the low exergy recovery and unmatched heat capacity, especially with respect to the latent heat between the heat of condensation of compressed steam and the heat of water evaporation from BL. A low target solid content means a low amount of water evaporating from the BL. Therefore, the amount of compressed steam that becomes a heat source during evaporation decreases accordingly, leading to largely unmatched amounts of latent heat during heat exchange inside the evaporator. Consequently, the compressor works harder and acts as a heater providing sensible heat to cover evaporation's heat. The smallest total work required can be achieved with a target solid's content of 75 wt% wb. The energy consumption increases slightly as the BL concentration is increased

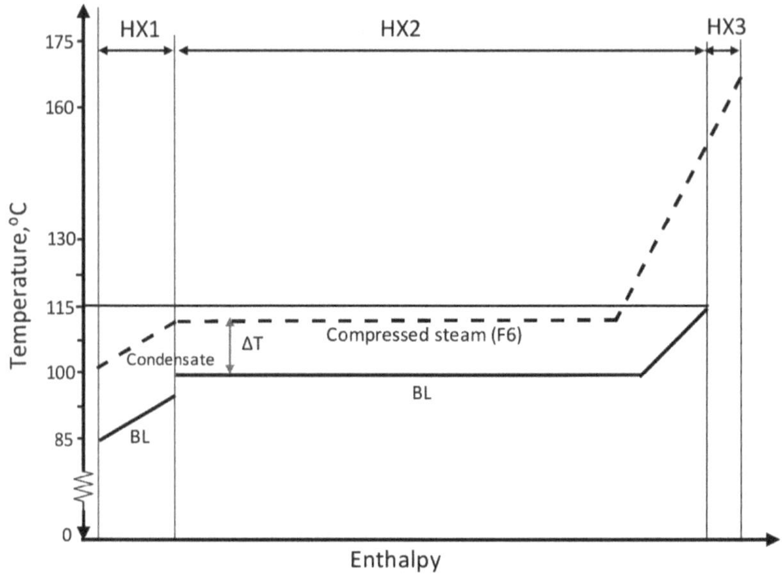

Figure 4.10 Temperature—enthalpy diagram during black liquor (BL) evaporation using self-heat recovery.

further. This slight increase is mainly due to an increase in the evaporating temperature, which consequently leads to an increase in the compressor work.

Fig. 4.10 indicates the temperature—enthalpy diagrams all through BL evaporation. In the calculation, the final water content of BL was determined to be 27.5 wt% wb (concentrated BL solid of 72.5 wt% wb). Therefore, based on Eq. (4.1), the specified evaporation temperature is 115°C. All energy required for the evaporation process (work), W_{tot}, is 8.1 MW, consisting of compressor and blower duties of 7.61 and 0.49 MW, respectively. Also, the outlet steam pressure and compressor temperatures are 155 kPa and 167°C, respectively. The solid and dashed lines indicate the hot and cold flows, respectively. Here, ΔT is the temperature difference between the water evaporation temperature of the BL and the compressed steam condensation temperature inside the evaporator (steam tube rotary evaporator) at 13°C. Latent heat exchange is dominant in the steam tube rotary evaporator and it occurs between the hot (compressed vapor) and cold (BL) streams. The compressed vapor curves are thus nearly parallel to those of the cold streams of BL water. This phenomenon demonstrates the excellent heat coupling for each type of heat, including sensible and latent heat. As a result, the loss of energy during BL evaporation could be minimized and the energy required for BL evaporation would significantly be reduced.

4.2.5 System's efficiency

Fig. 4.11 provides an overview of the work required during BL evaporation and the overall net energy (η_{net}) performance by using different BL evaporation technologies. In order to evaluate each system's performance, the solid content target was set at 72.5 wt% wb. The BL evaporation technologies assessed include conventional evaporation (boiler-based evaporation, boiler efficiency of 73%), MEE, and the proposed exergy-based BL evaporation approach. In general, BL evaporation using the suggested self-heat recovery method needs the least energy for the drying process, followed by MEE and conventional evaporation. The efficiency of the overall system that adopts conventional evaporation is about 14.7%. On the other hand, the system's efficiency if using MEE is 24.5%, which is around 10% higher than in the conventional scenario. In MEE's case, the water evaporated by live steam in the first stage is not immediately exhausted and disposed to the condenser. Instead, it is retained as a heat source in the second evaporation stage, thus reducing steam usage. Unfortunately, when BL is concentrated until its solid content exceeds around 50%, its complex behavior changes, resulting in liquor viscosity. As a result of this increase in viscosity, a high-solid concentrator is required by adopting direct-contact heat transfer; thus, additional high-temperature

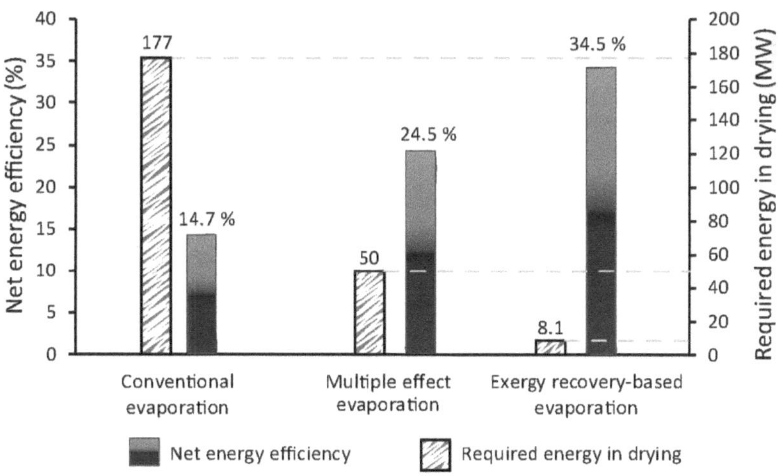

Figure 4.11 Comparison of the energy performance among black liquor (BL) evaporation technologies.

steam is needed. In the exergy recovery-based system, the evaporator system needs an energy input rate of 8.1 MW for evaporation, significantly lower than the other two evaporator systems' specifications. The evaporation dynamics within the evaporator, including heat transfer, can also be enhanced using this approach. The proposed integrated system could achieve an overall efficiency of 34.5%, considerably higher than that employ the MEE (24.5%) and conventional boiler-based evaporation (14.7%) systems.

4.3 Black liquor—based hydrogen and power coproduction combining supercritical water gasification (SCWG) and chemical looping

Because biomass-based fuel has to be relatively competitive with other energy sources, it is important to investigate effective biomass conversion routes. Biomass waste, such as BL from the pulp mill, has energy capacity for power and H_2 generation on an industrial scale. Proper and successful BL use will decrease environmental impacts and improve waste by optimizing both economic and energy values. However, since the water content of BL is very high, about 85 wt% wb, before converting or harvesting the energy from weak BL, drying becomes a necessary pretreatment. Unfortunately, because drying is an energy-intensive process, overall energy efficiency is theoretically reduced. To attain high total energy efficiency, a conversion method that can bypass drying is therefore favored. Supercritical water gasification (SCWG) has certain advantages among gasification technologies, such as relatively lower operating temperature and no drying requirement prior to gasification [12,13]. Gasification is also known to be a more efficient method with better performance than other methods in terms of material decomposition and the value of chemical energy.

Various routes of BL utilization have been proposed to generate electricity, biofuel, and heat on an industrial scale. At supercritical conditions of up to 470°C and 27 MPa, Huet et al. [14] conducted an experimental analysis of SCWG of BL. They found the gas generated from the process was predominantly a mixture of H_2, CH_4, and CO_2. At a temperature of 470°C, the optimum result could also be obtained. All the organic carbon was able to decompose at that temperature. However, their investigation did not require any attempt to minimize the loss of energy required to circulate and recover energy in the system. Cao et al. [12] researched and examined the H_2 generation by SCWG, the separation of H_2 using pressure swing adsorption, including steam turbine power generation. Significant loss of energy occurred during the process due to the energy-intensive process during separation. In addition, the effect of BL evaporation on energy and exergy has also been studied by Cao et al. [15]. They found that by implementing BL evaporation before SCWG, the overall efficiency of the system decreased. Andersson and Harvey [10] reported the efficiency of the conventional BL gasification process to produce H_2, focusing on the emission of CO_2.

4.3.1 General conditions and detailed proposed process

The proposed integrated system is shown in Fig. 4.12, representing a detailed flow diagram for energy-efficient coproduction of BL to power and H_2. During SCWG, parameters are shown and summarized in Table 4.4 [16]. The SCWG is considered the key conversion method due to its excellent characteristics, such as no drying requirements and the ability to produce H_2-rich syngas. The SCWG reactor is initially fed with weak BL with high water content, and water is supplied and used as the reactant. To be remembered, slurry weak BL is pumped and it is preheated (HX1 and HX3) and flown to the SCWG reactor at the target pressure for SCWG (higher than 22 MPa).

Due to its ability to avoid plugging, high gasification efficiency, and continuous production of syngas with high conversion rates, a fluidized bed

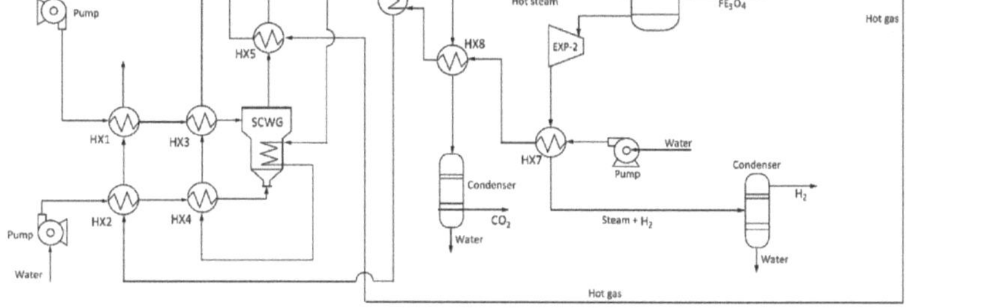

Figure 4.12 Schematic diagram of supercritical water gasification (SCWG) of black liquor (BL), syngas chemical looping (SCL), and power generation.

Table 4.4 Main conditions during supercritical water gasification (SCWG).

Specification	Value
Gasification temperature (°C)	375, 500 and 650
Pressure condition (MPa)	40
Type of gasification	Supercritical water gasification
The flow rate of solid BL (t/h)	50.4
Water content (%)	90

is adopted as an SCWG reactor [17]. For optimal heat transfer and temperature distribution, the silica sand is inserted inside the reactor as fluidizing particles. On the other hand, additional water is used as the fluidization medium fed from the bed bottom. Formulas [18] are used to estimate the pressure drop and the minimum bubbling velocity around the fluidized bed. Inside the gasifier, heat exchanger tubes are also mounted. The water and hydrogen bond density decrease considerably under the supercritical condition. Water can act as a nonpolar solvent as a result. Therefore, under homogeneous conditions, the reaction can be conducted [19].

In the ASPEN Plus process simulator, the RGibbs reactor and heat exchangers represent the gasification and superheating processes, respectively. The primary objective of the simulation is to examine the heating required to ensure the SCWG operation. Although the performance of syngas is extracted from other experimental studies [16], the SCWG overall exergy balance mentioned here can be written as,

$$\dot{E}_{BL} + \dot{E}_{water} + \dot{W}_{el} = \dot{E}_{gas} + \dot{E}_{waste\ heat} + \dot{I}_{overall} \tag{4.2}$$

Where $\dot{E}_{BL}$ and $\dot{E}_{water}$ are the exergy rate of the BL and fluidizing water, respectively, $\dot{W}_{el}$ represents the electric power supply, $\dot{E}_{gas}$ is the exergy rate of the produced gas, $\dot{E}_{waste\ heat}$ is the external exergy loss from the condenser (unrecovered heat), and $\dot{I}_{overall}$ represents the overall internal exergy loss in the system (irreversibility rate). Since SCWG is a complicated process consisting of many chemical reactions, there are two main reactions during gas production [20] as follows:

$$CH_xO_y + (2 - y)\ H_2O \rightarrow CO_2 + (2 - y + 0.5X)H_2 \tag{4.3}$$

$$CH_xO_y + (1 - y)\ H_2O \rightarrow CO + (2 - y + 0.5X)H_2 \tag{4.4}$$

Where x and y are H/C and processed BL molar ratios of O/C, respectively. Before being returned and used as the heat source for

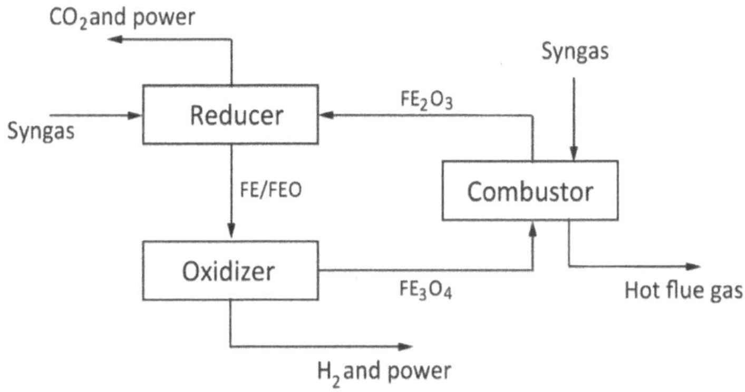

Figure 4.13 Conceptual diagram of syngas chemical looping (SCL) stage.

subsequent SCWG and preheating, the hot mixture of generated syngas and steam exhausted from the top of the SCWG reactor is then superheated (HX5) to increase its exergy rate.

Based on the cyclic reduction and oxidation processes, pure H_2 and concentrated CO_2 can be generated in two separate reactors within the SCL process. As conceptually shown in Fig. 4.13, the SCL mechanism consists primarily of three interconnected reactors: reducer, oxidizer, and combustor. In ASPEN Plus, using the RStoic reactor model, these processes are modeled. Based on the pilot-scale SCL plant currently in operation, this study uses a counter-current moving bed reactor for the reducer and oxidizer, while an entrained bed is used for the combustor [21]. In order to facilitate multiple reactions during the process, an iron-based material is used for the oxygen carriers (OCs). The solid mass fraction used in the SCL process is set at 70% Fe2O3, 15% SiC, and 15% Al2O3. The equilibrium gas concentration is also assumed during calculations in this report, based on preliminary investigation.

Due to higher kinetics conditions, the higher pressure leads to optimum gas–solid conversion. In order to achieve optimal performance and better efficiency during the reduction process, a high temperature of up to 900°C is recommended; hence, the syngas can be fully transformed to CO_2 and steam. In addition, to preheat the water, the heat from steam and CO_2 exhausted from the reduction reactor is recovered (HX-8). Afterward, in EXP-1, it is utilized to generate electricity via the expander. The reactions that occur during the process of reduction are given as follows:

$$Fe_2O_3 + H_2 \rightarrow 2FeO + H_2O \ldots \Delta H = 38.4 \text{ kj/mol} \qquad (4.5)$$

$$Fe_2O_3 + CO \rightarrow 2FeO + CO_2 \ldots \Delta H = -2.8 \text{ kj/mol} \quad (4.6)$$

$$FeO + H_2 \rightarrow Fe + H_2O \ldots \Delta H = 30.2 \text{ kj/mol} \quad (4.7)$$

$$FeO + CO \rightarrow Fe + CO_2 \ldots \Delta H = -11 \text{ kj/mol} \quad (4.8)$$

$$4Fe_2O_3 + 3CH_4 \rightarrow 8Fe + 3CO_2 + 6H_2O \ldots \Delta H = 897.175 \text{ kj/mol} \quad (4.9)$$

The reduced OCs react to the steam and generating H_2 in the oxidation phase, and it flows out along with the excess steam. In EXP-2, the stream containing high-pressure steam and H_2 leaving the oxidizer produces power via expansion. Once the steam is concentrated, high-purity H_2 can be fully obtained. The following are the oxidation reactions:

$$Fe + H_2O(g) \rightarrow FeO + H_2(g) \ldots \Delta H = -30.2 \text{ kj/mol} \quad (4.10)$$

$$3FeO + H_2O(g) \rightarrow Fe_3O_4 + H_2 \ldots \Delta H = -60.6 \text{ kj/mol} \quad (4.11)$$

The Fe_3O_4 produced in the oxidizer reacts in the combustor with oxygen; therefore, it is recycled back to Fe_2O_3. Syngas is also added and reacted in the combustor with O_2 to satisfy the SCL process's heat. To help the overall SCL process, this reaction provides more heat. Also, the heat from the combustor brought by the hot exhaust gas is used to elevate the air's exergy rate inlet stream. The remaining energy (heat and pressure) is, on the other hand, recovered by a power generation expander (EXP-3). Noticeably, as demonstrated in other research, the combustor has a pressure of 0.2 MPa higher than one in both the reducer and the oxidizer [22,23]. Here, the reactions involved are:

$$4Fe_3O_4 + O_2 \rightarrow 6Fe_2O_3 \ldots \Delta H = 471.6 \text{ kj/mol} \quad (4.12)$$

$$2H_2 + O_2 \rightarrow 2H_2O \ldots \Delta H = -457 \text{ kj/mol} \quad (4.13)$$

The detailed conditions during the SCL stage are summarized in Table 4.5.

4.3.2 Performance of integrated system and analyses

The calculation to evaluate the total efficiency of the proposed system is shown in Eq. (4.14):

$$\eta_{net} = \frac{P_{output} - P_{internal}}{P_{input}} \quad (4.14)$$

where the sum of generated H_2 and electricity, consumed power by the system, and total energy input from BL (MW), respectively, are P_{output}, $P_{internal}$, and P_{input}. The system's internal energy consumption consists of

Table 4.5 Assumed operating conditions during the syngas chemical looping (SCL) process.

Parameters	Value
Minimum temperature during reduction (°C)	900
Minimum temperature during oxidation (°C)	820
Minimum combustion temperature (°C)	1100
Pressure condition (MPa)	2–3.5
Isentropic efficiency of compressors (%)	90
Pumps efficiency (%)	90
Mass fraction of solid material	70% Fe_2O_3, 15% Al_2O_3, 15% SiC
Oxidation and reduction reactor	Counter-current moving bed
Combustion reactor	Entrained bed

the pump duties before the SCWG and the SCL compressors and pumps. In this analysis, under several SCWG conditions and operating pressures of the SCL reactors, the developed system is examined to assess the system's energy efficiency performance.

Fig. 4.14 demonstrates the influence of the gasification temperature on the produced gas during the SCWG operation and on the total H_2

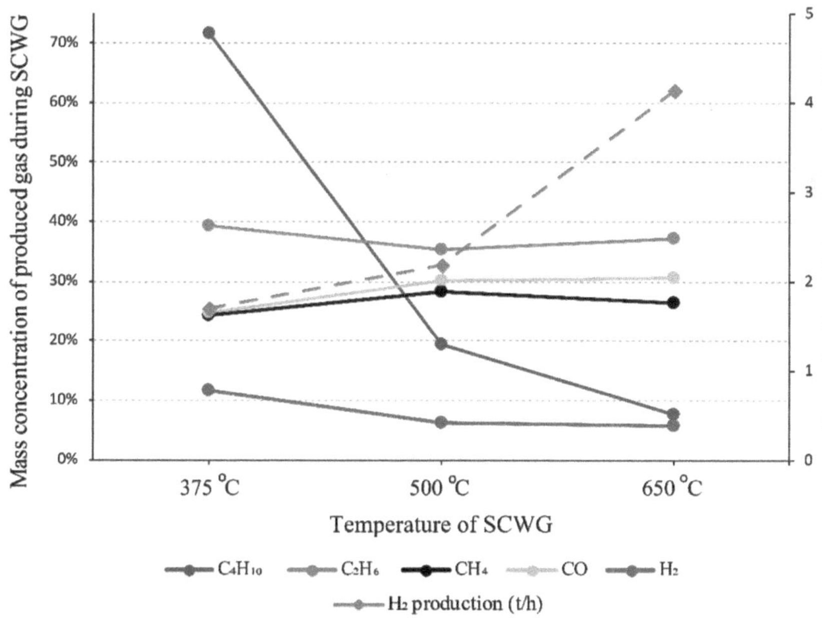

Figure 4.14 The effect of supercritical water gasification (SCWG) temperature on the mass concentration of the gas produced and final H_2 generated.

generated. The result shows that SCWG at 650°C leads to higher production of H_2, which is 4.13 t/h. Increased SCWG temperature generally leads to higher production of H_2 during the SCL process. In addition, although the H_2 concentration of the gas produced during gasification is lower at higher gasification temperatures, the H_2 concentration is still higher than at lower temperatures. After gasification, CO, CH_4, and C_2H_6 are oxidized to produce CO_2, while OCs are reduced to Fe and FeO. Since the syngas composition is estimated based on a laboratory scale, it is important to validate it on the pilot scale. It can also observe the high-pressure fluidization behavior at this level, such as the pressure drop and the minimum bubbling velocity across the fluidized bed reactor, not only to quantify the gas production during the SCWG process.

Fig. 4.15 illustrates the effects of temperature during SCWG on SCL output, such as H_2 formed, CO_2 extracted, and overall system efficiency. The system performance significantly increased in higher temperatures. The outcome shows that the SCWG carried out at 650°C resulted in around 80% of the highest process efficiency. When the SCWG temperature changed from 375 to 650°C, the amount of CO_2 captured increased from about 11 to 28.35 t/h. The captured CO_2 corresponds to a 75% CO_2 capture efficiency. During the syngas combustion, the remaining CO_2 is released into the atmosphere to provide heat for the SCL process. It is not necessary to remove the water content beforehand in the overall integrated BL recovery scheme. Also, 75% of the concentrated CO_2 stream can be

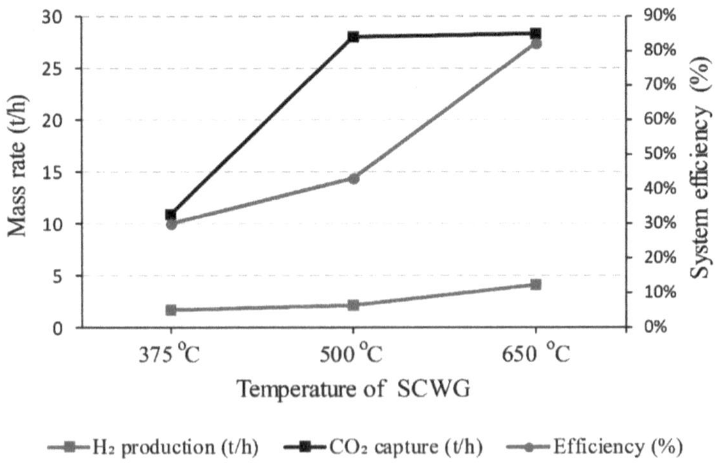

Figure 4.15 Correlation of temperature during gasification with syngas chemical looping (SCL) performance.

condensed after the SCL stage's oxidation phase, avoiding CO_2 separation energy requirements. A highly effective and environmentally cleaner BL utilization system leads to efficient SCWG and SCL processes implementing exergy-recovery and process integration technologies.

4.4 Efficient black liquorcogeneration of hydrogen and electricity via gasification and syngas chemical looping

Hydrogen (H_2) has been recognized as an attractive energy carrier of renewable energy in the future due to its characteristics such as high performance, diverse potential technologies for development and usage, and cleanliness [23]. Therefore, both production and effective utilization technologies are being actively developed, both on a large and small scale. Like electricity, H_2 theoretically contributes to zero carbon emissions as an effective energy carrier (secondary energy source) at the point of use [24]. Unfortunately, hydrogen is largely available on earth in an oxidized state (water), although it has high specific energy content. H_2 should therefore be produced using appropriate conversion technology, such as gas reforming, oil reforming, coal and biomass gasification, and electrolysis splitting of water. Gasification is considered a well-established technology compared to other methods, providing better results regarding increases in material decomposition and chemical energy.

An alternative technology with a faster reaction rate and excellent conversion efficiency is believed to be BL conversion by gasification. This approach can replace the inefficient and costly conventional recovery process due to the high BL water content, which varies between 83% and 87% on a wet basis. Conventional recovery boilers demonstrate a very low energy efficiency of about 9%–14% in electricity generation. Naqvi et al. [25,26] found that O_2-blown gasification resulted in higher conversion efficiency (cold gas efficiency of 58%) than air-blown gasification. By employing O_2 as the gasifying agent, syngas dilution with nitrogen can be avoided as well. Unfortunately, their study did not focus on the effort involved to sufficiently circulate and recover energy throughout the system. The combination of BL gasification and the combined cycle (BLGCC) system with and without CO_2 capture was investigated by Ferreira et al. [8], focusing on exergy and economic analyses. Darmawan et al. [27] have recently proposed an integrated system for the production of electricity from BL adopting entrained flow gasifier.

4.4.1 The overall proposed cogeneration system

A schematic representation of the overall cogeneration plant of H_2 and electricity from BL can be seen in Fig. 4.16. In the next section, comprehensive information is provided about heat recovery and process integration. BL drying, circulating fluidized bed (CFB) gasification, SCL, and power generation are primarily part of the overall system. The solid, dashed, and dotted lines reflect, respectively, material, heat, and electricity.

In order to remove water from weak BL, the BL drying process is performed. Due to its favorable aspects, such as high heat transfer, good thermal performance, good handling capacity, and suitability for continuous operation, this method uses a steam tube rotary dryer with an internal heat exchanger. The concentrated BL is transferred to the CFB gasifier after being dried for the conversion of syngas. Syngas is generated sequentially inside the CFB gasifier. Since there are large quantities of high-temperature water in the hot syngas, heat recovery is used and combined with the SCL process to produce steam for power generation. Syngas is also purified from particulate matter, including sodium carbonate (Na_2CO_3) and H_2S. Three reactors are adopted by the SCL process, namely the reducer, oxidizer, and combustor. OCs react with the syngas generated in the CFB gasifier, producing CO_2 and H_2O in the reducer. In the oxidation of OCs, iron oxide, by steam in the oxidizer, H_2 is produced. By combustion with air, the OCs are then recovered to their initial state. By using expanders to produce power, the heat produced in each reactor is recovered.

Assuming that the BL evaporation technique was explained earlier in Section 4.2, only the CFB gasification and SCL integration process is seen in Fig. 4.17. To produce relatively high-temperature oxidation steam, the generated syngas leaving the gasifier is recovered by the heat exchanger (HX-5). The syngas is compressed to a pressure of 2–3.5 MPa after

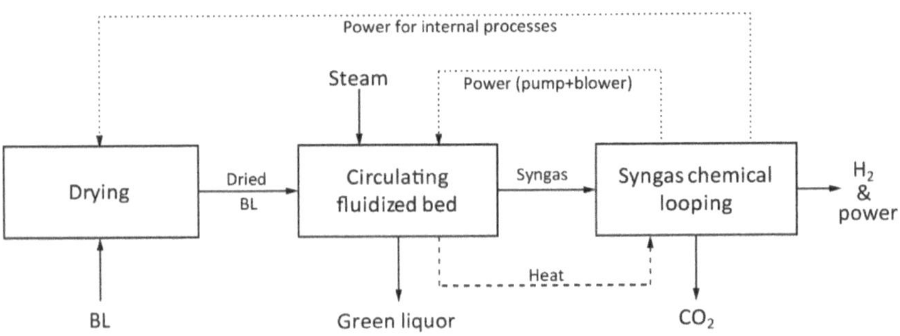

Figure 4.16 Conceptual diagram of the coproduction system of H_2 and power.

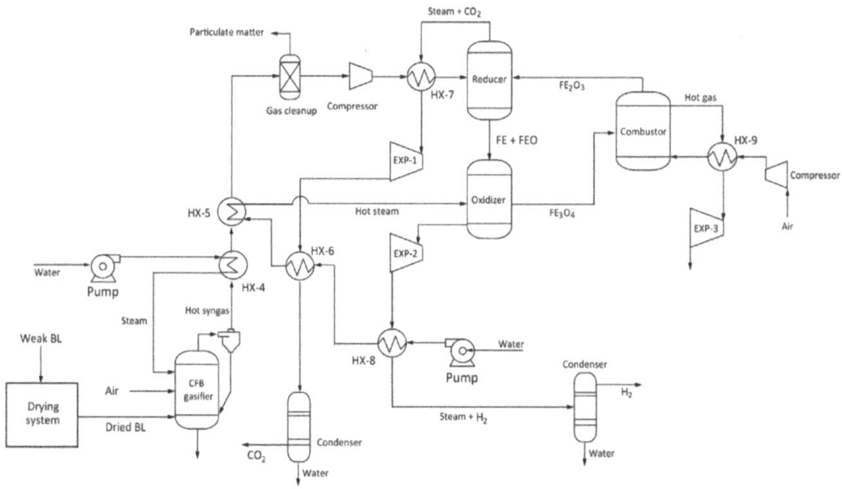

Figure 4.17 Process flow diagram of gasification and syngas chemical looping (SCL).

cleaning to increase its exergy level. The steam and CO_2 generated in the reducer are subsequently used to preheat the syngas (HX-7), and power is produced in the expander (EXP-1); H_2O (HX-6) is then preheated before it flows into the condenser. The generated stream containing steam and high-pressure H_2 from the oxidizer (EXP-2) is used for power generation. Notably, the reactor works at 0.2 MPa higher in the combustor than in the reducer and oxidizer. In addition, to increase the air intake stream's exergy rate, the heat carried by the hot gas from the combustor is used and the remaining energy is recovered to generate electricity (EXP-3).

The total net energy efficiency of the integrated system is calculated as follows:

$$\eta_{net} = \frac{P_{output} - P_{internal}}{P_{input}} \tag{4.15}$$

where P_{output}, $P_{internal}$, and P_{input} are total produced H_2 and generated power, internal power consumption, and total energy input (the BL input (MW)), respectively. Internal energy usage consists of a compressor and a blower in the gasifier for drying, pumping, and compressor duty, and SCL for compressor and pumps work. In addition, to analyze the effects of different operating parameters on system energy performance, the integrated cogeneration system is simulated and assessed at different steam-to-fuel ratios during the gasification and operating pressures of the SCL reactors.

4.4.2 Process modeling and calculation of gasification and syngas chemical looping

The CFB gasifier operates direct gasification in this analysis, where the fuel is partially oxidized to provide the heat needed for the operation. The fluidized bed gasifier usually works in the 750–850°C temperature range. The steam-to-fuel ratio or air-to-fuel ratio also affects this temperature. The CFB gasifier has higher carbon conversion efficiency and excellent heat transfer capacity than the conventional bubble fluidized bed gasifier [28,29]. After drying, the concentrated solid BL (80 wt.% wb) is gasified to generate H_2, CO, and CH_4-containing syngas at high temperatures. Other hydrocarbons and tar species are estimated to generate in the gasifier. However, since BL contains large quantities of catalytically active Na, the content of hydrocarbons is lower than that of other biomass (e.g., wood) [30,31]. Syngas, which, due to water's presence, has considerable thermal energy, is used for steam production for gasification. The temperature of the syngas before cleaning is kept above 300°C to prevent gas condensation.

In Fig. 4.18 and Table 4.6, respectively, descriptions of process simulation and CFB gasification parameters are shown and summarized. The BL decomposition and gasification reactor are simulated as the RYield and RGibbs blocks, respectively, in the Aspen Plus simulator. The combustor stage adopts the block of RGibbs to reacting the carbon in BL with oxygen. Since BL is known as a nonconventional solid, other reactants cannot react directly to it. To break down BL solid into H_2O, ash, and components, such as O_2, H_2, and biomass, an RYield block with a FORTRAN

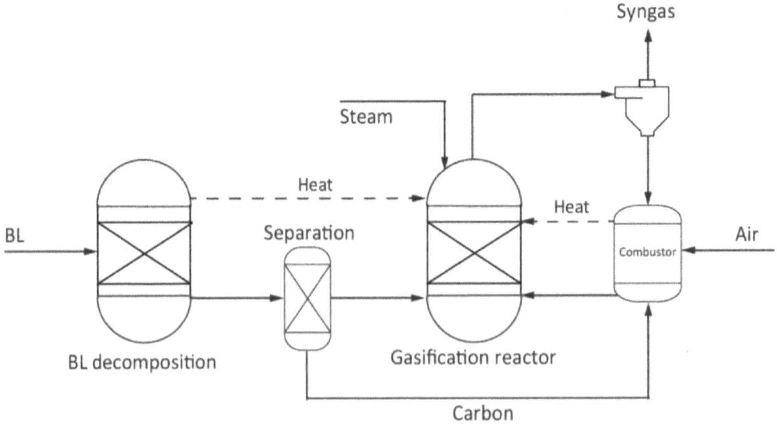

Figure 4.18 Process simulation of circulating fluidized bed (CFB) gasification in Aspen Plus.

Table 4.6 Assumed conditions during circulating fluidized bed (CFB) gasification.

Parameters	Value
Gasifier	
Temperature (°C)	800
Gasifier pressure (MPa)	0.1
Gasifier type	Circulating fluidized bed
Gasifying medium	Steam-air
Steam-to-fuel ratio	0.1–0.8
BL moisture content (wt.% wb)	20

subroutine is then used. In order to provide the heat for gasification, a portion of solid carbon is reacted with air. The remaining decomposed components are then introduced into an RGibbs block to facilitate the desired chemical reactions. The reaction heat for BL conversion is equal to the sum of the RYield (BL decomposition) block, RGibbs (gasification reaction) block, and RGibbs (combustor).

4.4.3 Syngas chemical looping and power generation system

Cyclical treatment of the iron oxide–based looping medium with syngas and steam is carried out in the SCL module to coproduce H_2 and power from syngas. The SCL process consists primarily of reduction, oxidation, and combustion in interconnected reactors to generate H_2, CO_2, and electricity. For the oxidation and reduction reactors, a counter-current moving bed reactor is adopted, whereas, for the combustion reactor, an entrained bed is used. Iron-based OCs are used in this process to enhance the reactions of reduction and oxidation. The mass fraction of the solids circulated in the SCL system is assumed to be 70% Fe_2O_3, 15% SiC, and 15% Al_2O_3.

Syngas is introduced into the reduction reactor after gas cleaning and is reacted with O_2 carried by Fe_2O_3. A high pressure (up to 3.5 MPa) is suggested at this reduction stage to achieve the optimum gas–solid conversion and increase the kinetics that leads to a smaller size of the reactor [21,32]. To achieve excellent performance, the reduction process is performed at a minimum temperature of 900°C, and the syngas is completely transformed to steam and CO_2. Depending on the particle reduction rate and the syngas' composition, the reducer's overall reaction may be either slightly exothermic or endothermic. The reduction temperature was set to 930° C in this study to examine the performance of the system. The reactions that occur in the reducer are given as follows:

$$Fe_2O_3 + H_2 \rightarrow 2FeO + H_2O \quad (4.16)$$

$$Fe_2O_3 + CO \rightarrow 2FeO + CO_2 \quad (4.17)$$

$$FeO + H_2 \rightarrow Fe + H_2O \quad (4.18)$$

$$FeO + CO \rightarrow Fe + CO_2 \quad (4.19)$$

$$4Fe_2O_3 + 3CH_4 \rightarrow 8Fe + 3CO_2 + 6H_2O \quad (4.20)$$

The reduced particles react with steam at the oxidation stage to generate an H_2-rich gas stream that flows out with the remaining steam. A high-purity H2 stream can be extracted until the steam is condensed out of the gaseous state. In this process, the oxidation reactions involved are as follows:

$$Fe + H_2O(g) \rightarrow FeO + H_2(g) \quad (4.21)$$

$$3FeO + H_2O(g) \rightarrow Fe_3O_4 + H_2 \quad (4.22)$$

The Fe_3O_4 formed in the oxidation reactor is reacted in the combustor with O_2 in order to convert it back to Fe_2O_3. This exothermic reaction also gives the heat for the reduction reaction:

$$4Fe_3O_4 + O_2 \rightarrow 6Fe_2O_3 \quad (4.23)$$

The assumptions and conditions associated with the SCL process are given in Table 4.7.

4.4.4 General assumptions

The Aspen Plus V8.8 (Aspen Technology, Inc.) process simulator is used to model and perform calculations associated with material and energy balances in the proposed integrated system. The flow rate of weak BL entering

Table 4.7 Assumptions and conditions associated with the syngas chemical looping (SCL) process.

Parameters	Value
Reducer temperature (°C)	930
Oxidizer temperature (°C)	820
Combustor temperature (°C)	1000
SCL pressure condition (MPa)	2–3.5
Compressor isentropic efficiency (%)	90
Pump efficiency (%)	85
Mass fraction of circulated solid material	70% Fe_2O_3, 15% Al_2O_3, 15% SiC

Table 4.8 Composition of black liquor (BL) used herein.

Properties	Value
Total solids (wt.% wb)	15
Water content (wt.% wb)	85
Components (wt.% db) [11]	C: 27.50; H: 3.75; O: 39.35; N: 0.07; Cl: 0.16; Na: 19.85; K: 3.12, S: 6.20
LHV (MJ kg^{-1}, db)	12.4

the drying module was 348.12 t/h (730 t/d pulp production taken into account). Also, a few extra assumptions were made. For modeling and performing calculations associated with material and energy balances in the proposed integrated system, the Aspen Plus V8.8 (Aspen Technology, Inc.) process simulator was used. The flow rate of weak BL entering the drying module was 348.12 t/h (considering pulp production of 730 t/d). Furthermore, a few additional assumptions were made: (i) the minimum temperature difference in all heat exchangers 30 K; (ii) the ambient temperature was set to 25°C; (iii) the adiabatic performance of the compressor and the blower was set to 90%; (iv) no heat loss; (v) no air leakage inside the dryer; and (vi) the dryer consist of a mixer, heat exchanger, and separator. Also, the air used was considered to contain 79 mol.% N2 and 21 mol.% O_2. Table 4.8 lists the composition of the BL used during the calculation.

4.4.5 Performance of gasification, syngas chemical looping system, and overall system

Fig. 4.19 denotes the steam-to-fuel ratio's effect on the composition of syngas and cold gas efficiency. The latter indicates the ratio between energy

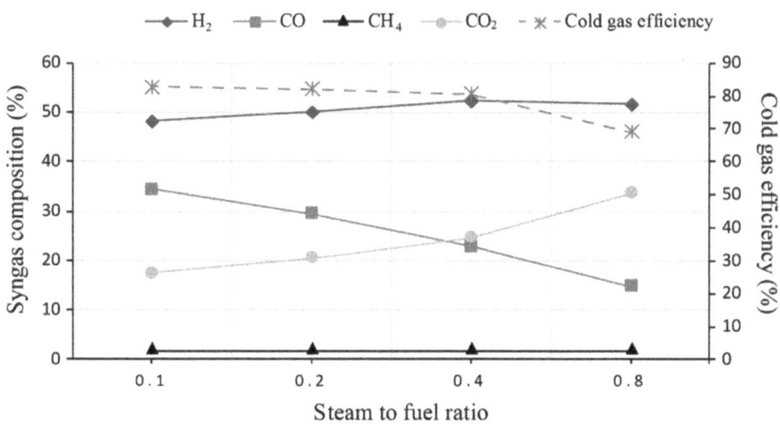

Figure 4.19 Effect of steam/fuel ratio during gasification.

flow in the syngas and the chemical energy stored within the BL. Increasing the ratio of steam to fuel usually results in an increased output of H_2 and CO_2. Unfortunately, CO decreases by introducing a larger quantity of steam, reducing the cold gas's efficiency. To maintain the temperature at 800°C during gasification, a higher steam-to-fuel ratio is needed, which means that a larger amount of air should be supplied to provide more heat via carbon combustion in the reactor. It has resulted in increased concentration of CO_2 and reduced efficiency of gasification. Ju et al. [29] have experimentally explained this phenomenon. Their findings confirmed that the rising amount of air due to the steam gasifier's endothermic reaction would lower the useful gas content. It can be inferred that the quantity of air and steam significantly affects the efficiency of gasification.

Since the steam-to-fuel ratio in the gasifier significantly affects the generated syngas, it potentially affects the SCL's efficiency. Fig. 4.20 depicts the effect of the steam-to-fuel ratio on the output of H_2 and the net generated power at the SCL pressure of 3.5 MPa. The proportion increase of steam to fuel from 0.1 to 0.8 would decrease the overall net power from 8.57 to 3.3 MW. Also, the lower quantity of steam in the gasifier results in a higher exergy rate that can be recovered in HX-5 before the oxidation reactor reaches the steam. The steam supplied to the oxidation reactor increases and increased the quantity of energy produced due to high energy recovery (EXP-2). Suppose the steam-to-fuel ratio is higher than 0.4, the output of H_2 decreases. The decreased H_2 and CO in the syngas decreases

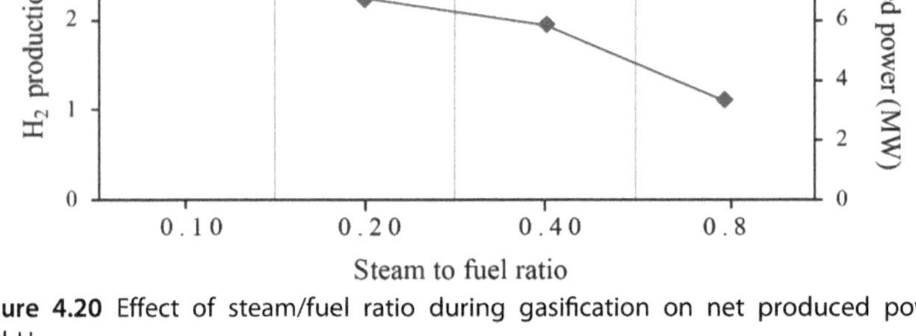

Figure 4.20 Effect of steam/fuel ratio during gasification on net produced power and H_2.

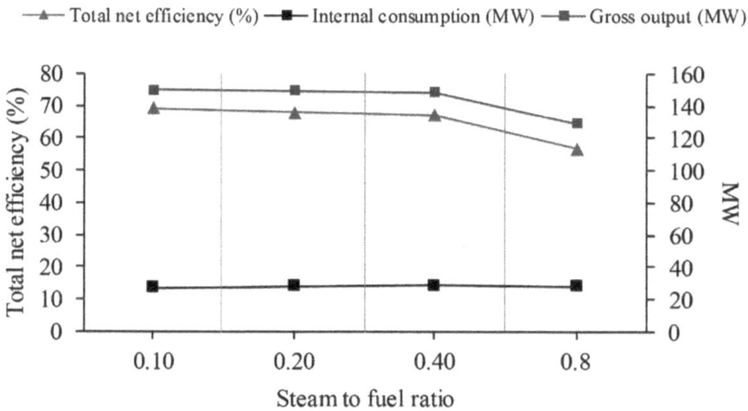

Figure 4.21 Correlation of steam/fuel ratio during gasification with net produced power and H_2 production.

the amount of Fe and FeO produced in the reducer, leading to a decrease in the overall H_2 yield.

The relation between the ratio of steam to fuel and the overall energy efficiency is presented in Fig. 4.21. Generally, in the gasification process, internal consumption is not greatly affected by additional steam. The internal energy consumption rises by 4.61% when the steam-to-fuel ratio is changed from 0.1 to 0.8. However, the rise in the steam-to-fuel ratio will substantially reduce the overall energy efficiency (power and H_2) from 69.16% to 56.74%. Compressor work (greater than 90%) dominates internal power consumption throughout the system, especially in the drying and SCL processes. Around 7.6 MW, or 28% of the total internal energy, is consumed by the drying system's compressor.

In Fig. 4.22, the efficiency of the integrated system at various SCL pressures is shown. The metal (Fe_2O_3 + inert) flow rate is set at 75 kg/s to properly analyze the effect of the SCL pressure condition on the process, with a fixed steam-to-fuel ratio of 0.1 during gasification. There is no significant effect on the overall net energy efficiency of the rise in the SCL operating pressure. However, after an increase in SCL pressure from 2 to 3.5 MPa, the net power generated increases slightly, which is observable in the slightly higher power generated in EXP-2.

In conclusion, by using process integration and energy recovery technologies, the proposed system will achieve a total net energy efficiency of almost 70%. In any process, these technologies can allow optimum use of unrecovered energy; hence, significantly high total energy efficiency can be

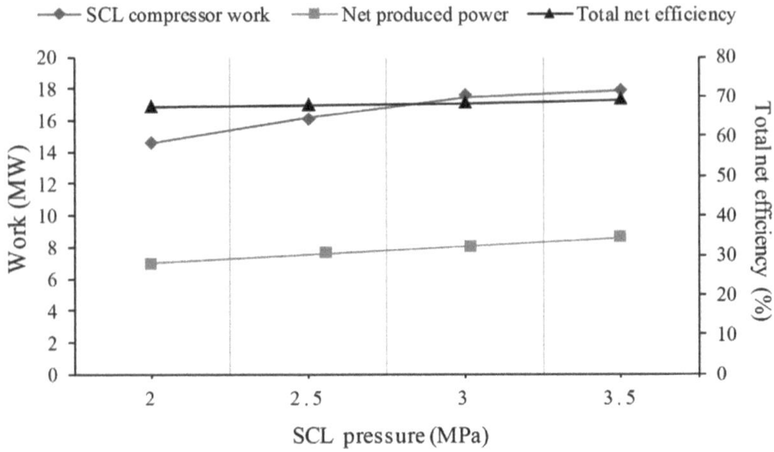

Figure 4.22 Effects of syngas chemical looping (SCL) pressure on system performance.

achieved. The integrated system, consisting of drying, gasification, and SCL, appears very promising compared with other BL recovery systems. Also, a concentrated CO_2 stream can be directly extracted during the SCL process, thereby avoiding additional CO_2 separation energy requirements. It is concluded that the proposed integrated system improves cleaner and more efficient BL utilization.

4.5 Coproduction of power and ammonia from black liquor

H_2 storage is very challenging work because of its very low volumetric energy density. Due to its high performance, high H_2 density, applicability and simple catalytic decomposition, ammonia (NH_3) is considered a very attractive method for H_2 storage [33,34]. Also, due to the well-developed technology of synthesis and distribution, it is considered high economic efficiency. NH_3 has the greatest density for H_2 storage relative to other H_2 storage media, such as methylcyclohexane, methanol, and ethanol (as much as 17.8 wt%). NH_3 is therefore commonly used in many fields, especially in the energy sector, such as fuel cells, direct engine combustion, heating, power generation, and even hybrid vehicles.

NH_3 synthesis is achieved primarily through the Haber–Bosch loop system, the electrochemical process, and the membrane reactors. However, the energy required for electrochemical processing is expected to be 20% lower than that of the Haber–Bosch process. Conversely, due to the high energy requirement to dissociate the triple-bonded nitrogen molecule, the Haber–Bosch process requires a high temperature (400–600°C) and a high pressure (up to 30 MPa). While both the electrochemical process and membrane reactors have adequate ability to reduce energy consumption during synthesis, despite the many efforts to improve these features, they are not mature enough for use as a practical technology [35]. The Haber–Bosch method, as we know, requires N_2 and H_2 as feedstocks to generate NH_3. Normally, N_2 is provided by cryogenic or other types of processes via air separation. Via oil and gas reforming, coal gasification, and water electrolysis, H_2 can be extracted from different materials, including natural gas and oil. Due to the high pressure and temperature needed, high energy input is demanded during NH_3 synthesis. During the processing of N_2, an additional energy penalty is also levied, which substantially decreases the overall energy efficiency [36]. Therefore, several experiments have been designed further to improve the energy efficiency of the conversion process. With the help of anhydrous $CaCl_2$, Malmali et al. [37] studied the possibility of NH_3 synthesis via a low-pressure reactive-absorption process. Nonetheless, further desorption process is required to release the NH_3 from $CaCl_2$, providing another energy penalty.

In addition, Haber–Bosch reactors, solid oxide electrolyzers (SOEs), and PSA-based N_2 production have been considered and evaluated for the combined production of NH_3 [38]. The technology established is capable of using the excess electricity generated from renewable energy sources using the SOE. A combined method that integrates steam methane reforming and the Haber–Bosch process was also examined by Frattini et al. [39]. Unfortunately, instead of making an effort to intensify and integrate each procedure, these studies merely merged multiple processes, resulting in low overall energy performance. Given the high potential of BL as an energy source and NH_3 as an efficient H_2 storage medium, an integrated high total energy efficiency system to convert BL to NH_3 is required. However, a search of the Scopus and Web of Science databases revealed that no such reports existed, particularly an integrated high energy efficiency system.

4.5.1 Overall process combination and common assumptions

This section develops a method involving BL evaporation, gasification, SCL, and NH_3 synthesis, as seen in Fig. 4.23, since the previous system did not cover the H_2 storage system. Exergy recovery and process integration approaches are combined to achieve efficient heat circulation in order to achieve high overall energy efficiency.

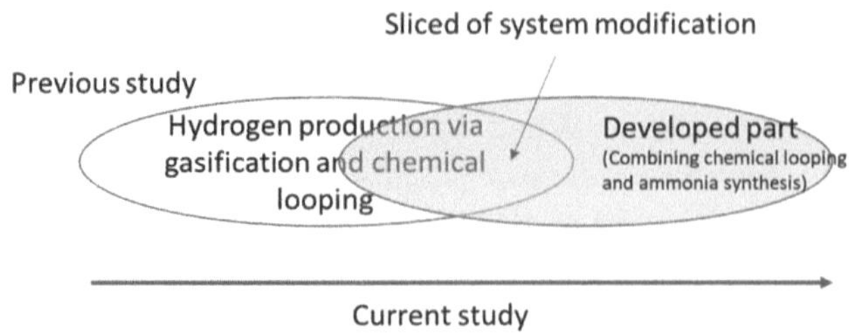

Figure 4.23 The current study covers hydrogen production and its storage with some modifications.

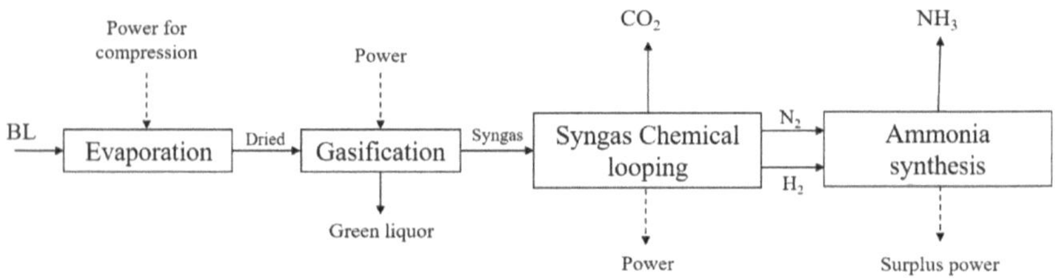

Figure 4.24 Schematic of the combined system of H_2 production and NH_3 synthesis.

The integrated conceptual system consists mainly of the production of syngas through gasification, the production of H_2 by SCL, and the storage of H_2 through NH_3 synthesis via the Haber–Bosch process. Considering the current average output in the pulp mill, the flow rate of BL entering the evaporation process is set at 348.12 t/h. Fig. 4.24 presents the general diagram of the process. Exergy recovery and process integration technologies were introduced throughout the process to minimize energy loss in the system. The recovery of energy provides intensification by circulating the energy/heat involved in each process [40]. Because of the limitation of heat recovery in a single process, process integration is adopted to recover other processes' excess heat. This technique decreases the loss of energy in the overall system and enhances total energy efficiency.

BL evaporation and CFB gasification are primarily used in the method for the production of syngas. Water extraction by evaporation is required prior to the gasification process at high temperatures. A steam tube rotary evaporator is adopted in the system due to the excellent heat transfer capacity, high thermal performance, and suitability for continuous processes [41]. Exergy recovery is conducted by recycling the steam to provide heat in the evaporation process. The heat recovery is also adopted by using the syngas generated from gasification containing high-temperature water to support the gasification process's heat requirement and provide steam for the SCL process. Also, particulates and other unwanted constituents, such as sodium carbonate (Na_2CO_3) and hydrogen sulfide, are removed from the syngas produced (H_2S).

Three key reactors are involved in the SCL process: the reducer, the oxidizer, and the combustor. Syngas reacts with OCs in the reducer to form CO_2 and steam. OCs leaving the reducer were subsequently added to the oxidizer and reacted with steam to produce H_2 with unconverted steam at a temperature range of 500–750°C. Therefore, by simply condensing the steam at this point, pure H_2 can be obtained, assuming that full condensation occurs. Afterward, the used OCs are restored by oxidization to their original state. In each phase, the produced hot gas is recovered by expanding it to produce electricity. Finally, it is beneficial to integrate NH_3 synthesis with the SCL process as both high-purity H_2 and N_2-rich gas are subsequently produced in the oxidizer and the combustor, respectively, without extra energy for gas separation.

4.5.2 Syngas chemical looping and NH_3 synthesis

The process flow diagram of the integrated system covers only the SCL process and NH_3 synthesis, as shown in Fig. 4.25 since evaporation and gasification were explained earlier. Pure H_2 and concentrated CO_2 can be produced in the SCL process in two separate units based on cyclic reduction and oxidation mechanisms, respectively. The SCL mechanism requires three integrated reactors: the reducer, the oxidizer, and the combustor. The reducer and oxidizer use a counter-current moving bed reactor in this analysis, while the combustor adopts an entrained bed. To enhance numerous reactions during the process, an iron-based material is used as the OC. The solid mass fraction used in the SCL process is set at 70% Fe_2O_3, 15% SiC, and 15% Al_2O_3. As described in the previous section, to produce high-temperature steam for the oxidation phase, the heat of the syngas (S1) exhausted from the gasifier is recovered in HX-5. The syngas is subsequently cleaned and then compressed to 2–3.5 MPa to increase the exergy rate and improve the SCL process. The higher pressure contributes to optimum gas–solid conversion and a smaller reactor size because of the more desirable kinetic state. In order to achieve excellent performance, a high temperature of up to 900°C is recommended, enabling the syngas to be fully converted to CO_2 and steam.

In addition, for preheating the syngas, the steam and CO_2 heat exhausted from the reducer is recovered (HX-7). Afterward, in EXP-1, the

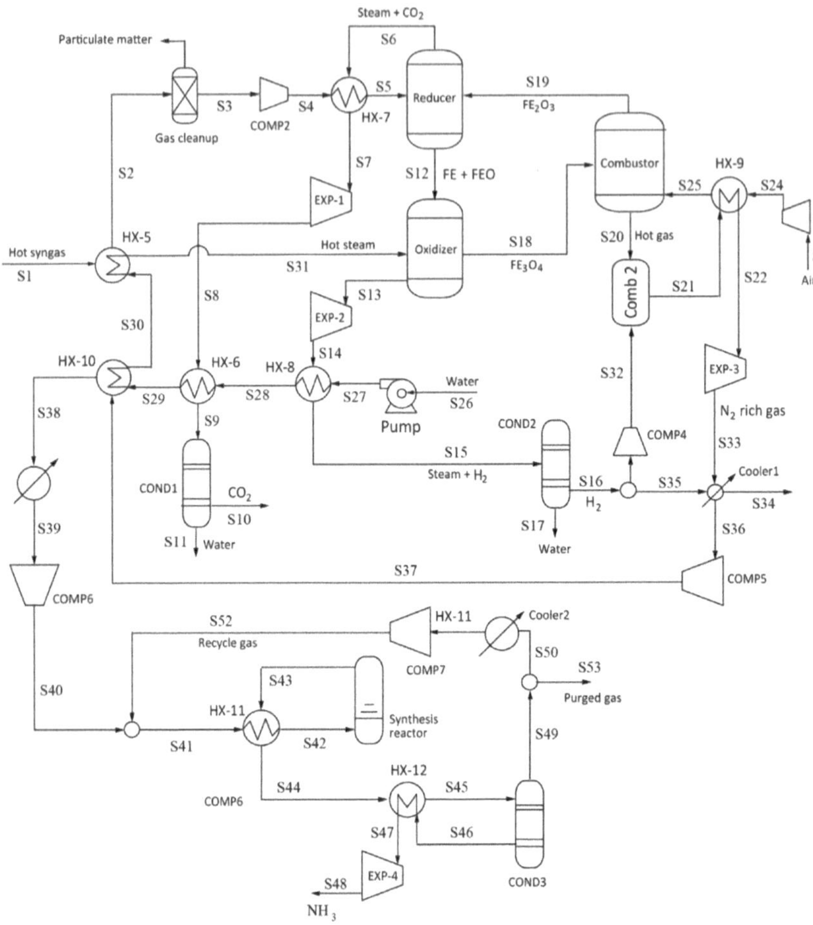

Figure 4.25 Process flow diagram of the syngas chemical looping (SCL) and NH_3 synthesis.

mixture is expanded to generate electricity. Also, in HX-6, H_2O is preheated before it reaches the condenser. The reactions during the process of reduction are given as follows:

$$Fe_2O_3 + H_2 \rightarrow 2FeO + H_2O \ldots \Delta H = 38.4 \text{ kj/mol} \tag{4.24}$$

$$Fe_2O_3 + CO \rightarrow 2FeO + CO_2 \ldots \Delta H = -2.8 \text{ kj/mol} \tag{4.25}$$

$$FeO + H_2 \rightarrow Fe + H_2O \ldots \Delta H = 30.2 \text{ kj/mol} \tag{4.26}$$

$$FeO + CO \rightarrow Fe + CO_2 \ldots \Delta H = -11 \text{ kj/mol} \tag{4.27}$$

$$4Fe_2O_3+3CH_4 \rightarrow 8Fe + 3CO_2+6H_2O \ldots \Delta H = 897.175 \text{ kj/mol} \quad (4.28)$$

The decreased OCs are reacted with steam to create H_2 in the oxidation stage, and H_2 flows out along with the excess steam. In EXP-2, the mixture of high-pressure steam and H_2 leaving the oxidizer produces power by expansion. Once the steam is concentrated, high-purity H_2 can be completely collected. The following are the oxidation reactions:

$$Fe + H_2O(g) \rightarrow FeO + H_2(g) \ldots \Delta H = -30.2 \text{ kj/mol} \quad (4.29)$$

$$3FeO + H_2O(g) \rightarrow Fe_3O_4 + H_2 \ldots \Delta H = -60.6 \text{ kj/mol} \quad (4.30)$$

Fe_3O_4, formed in the oxidizer, is reacted in the combustion chamber with oxygen and recycled back to Fe_2O_3. Before going into the NH_3 synthesis process, the amount of H_2 is introduced to the combustor (Comb 2) and reacted with O_2, forming H_2O. To help the overall SCL process, this reaction provides more heat. In addition, to increase the air intake stream's exergy rate, the heat carried by the hot gas from the combustor is used and the remaining energy is recovered to generate electricity (EXP-3). Notably, the combustor pressure is 0.2 MPa greater than that of both the reducer and the oxidizer. The reactions that are involved here are:

$$4Fe_3O_4 + O_2 \rightarrow 6Fe_2O_3 \ldots \Delta H = -471.6 \text{ kj/mol} \quad (4.31)$$

$$2H_2 + O_2 \rightarrow 2H_2O \ldots \Delta H = -457 \text{ kj/mol} \quad (4.32)$$

The detailed conditions during the SCL process are listed in Table 4.9.

It is beneficial to combine NH3 synthesis with the SCL process as both high-purity H_2 and N_2-rich gas are subsequently formed in the oxidizer and the combustor, respectively, without extra separation energy. In the synthesis reactor, NH_3 synthesis then occurs, and the mixture of generated NH3 and unreacted gas is used for preheating in the preheater (HX-12). It is used for generating power since the mixture has high pressure and temperature (EXP-4). In the condenser, the NH_3 and unreacted gas are then cooled and separated (COND3). The unreacted gas is subsequently split into the recycled stream (S50) and the purged gas (S53). To avoid the buildup of impurities within the stream and during the process, the gas must be purged. Notice that H_2, N_2, and a small quantity of NH_3 also comprise the purged stream. During NH_3 synthesis, the reaction is as follows (Table 4.9):

$$N_2+3H_2 \rightarrow 2NH_3 \quad (4.33)$$

Table 4.9 Conditions associated with syngas chemical looping (SCL) process and NH_3 synthesis.

Specifications	Value
SCL Process	
Minimum temperature of reduction (°C)	900
Minimum temperature of oxidization (°C)	750
Operating pressure (MPa)	2–3.5
Compressor adiabatic efficiency (%)	90
Pump efficiency (%)	85
Solid materials (in mass fractions)	70% Fe_2O_3, 15% Al_2O_3, 15% SiC
NH_3 synthesis	
Operating pressure (MPa)	15
Operating temperature (°C)	450
Catalyst	$Fe_{1-x}O$
Conversion rate per pass (%)	30
Purged stream ratio	0.1, 0.2, 0.3, 0.4

4.5.3 Analysis of energy performance

The total energy efficiency of the integrated system is calculated as follows when assessing system performance:

$$\eta_{net} = \frac{P_{output} - P_{internal}}{P_{input}} \tag{4.34}$$

where P_{output}, $P_{internal}$, and P_{input} are the total produced NH_3 and electricity, internally consumed electricity, and total energy input (the sum of BL input in LHV and additional input power), respectively. The system's power consumption consists of compression and blower operations in the gasifier for evaporation, pumping and compression operations, and SCL compression and pumping operations. To assess system energy performance, the established system is tested under the various operating pressures of the SCL reactors.

4.5.4 Calculation result and analyses

Fig. 4.26 illustrates the constructed network of heat exchangers corresponds to the SCL process and the NH_3 synthesis. In addition, Fig. 4.27 shows the integrated system's output at various SCL pressures. Although the increase in pressure during SCL affects the residence time inside the reactors, it does

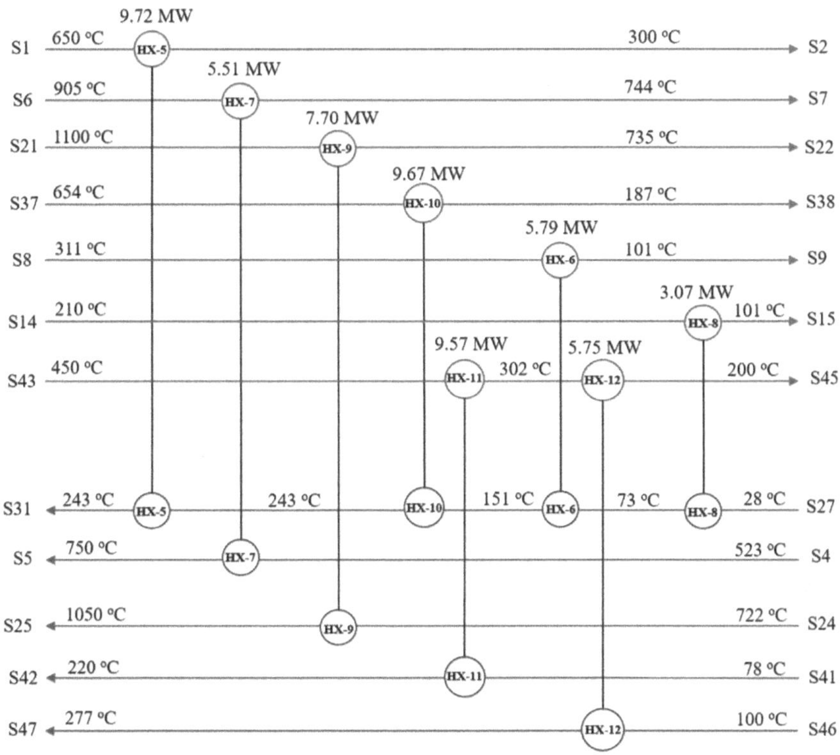

Figure 4.26 Heat exchanger network of syngas chemical looping (SCL) process and NH_3 synthesis (SCL pressure: 3.5 MPa).

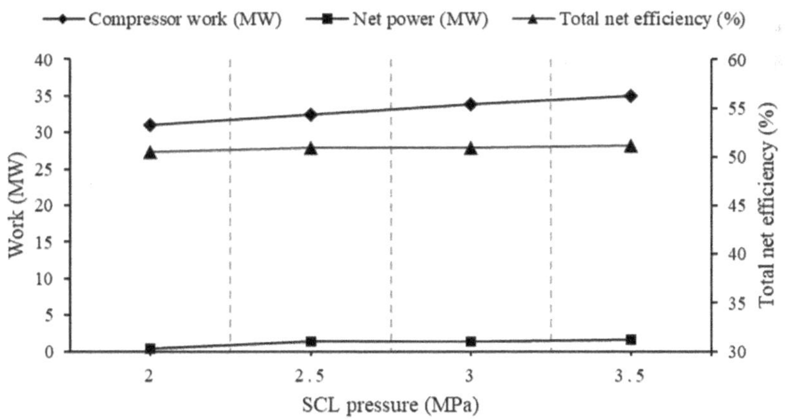

Figure 4.27 Impacts of syngas chemical looping (SCL) pressure on the system performance of SCL and NH_3 synthesis and overall efficiency.

not affect the integrated system's performance. However, the net power increases slightly during SCL and NH_3 synthesis from 0.39 to 1.60 MW following an increase in SCL pressure from 2 to 3.5 MPa due to the higher power produced. The integrated system's overall efficiency also slightly increased from 50.48% to 51.13%, while the output of NH_3 remained at around 18.14 t/h. When the purged stream ratio is raised to 0.4, it would significantly decrease the amount of NH_3 produced during the Haber–Bosch process to 13.07 t/h.

It is concluded that the result of the simulation suggests that N_2-rich gas and pure H_2 produced during SCL can be used directly without any energy penalty for NH_3 synthesis. It is also possible to eliminate the extra separation step for CO_2 capture, making the system cleaner and more efficient. Overall, with a total efficiency of about 50% for NH_3 and 100% for carbon capture, the proposed system is very promising. Since the current technologies for N_2 and NH_3 development consume large amounts of energy, we have implemented the SCL process and NH_3 synthesis in particular. In the future, the analysis might emphasize economic feasibility, comparison with current technology for the development of NH_3 (including catalyst types), and materials for the chemical looping process, including fuel types.

References

[1] Bajpai P. Black liquor gasification. 2014. https://doi.org/10.1016/B978-0-08-100009-0.00002-5.

[2] Zhang Y, Yao M, Gao S, Sun G, Xu G. Reactivity and kinetics for steam gasification of petroleum coke blended with black liquor in a micro fluidized bed. Appl Energy 2015;160:820–8. https://doi.org/10.1016/j.apenergy.2015.01.009.

[3] Bajpai P. Pulp and paper industry, vol. 1; 2016. https://doi.org/10.1017/CBO9781107415324.004.

[4] Naqvi M, Yan J, Dahlquist E. Black liquor gasification integrated in pulp and paper mills: a critical review. Bioresour Technol 2010;101:8001–15. https://doi.org/10.1016/j.biortech.2010.05.013.

[5] ÅF-Engineering. Energy consumption in the pulp and paper industry. Model Mills 2010 2011;1:53. doi:ÅForsk Reference: 09-163.

[6] Karlsson E, Gourdon M, Olausson L, Vamling L. Heat transfer for falling film evaporation of black liquor up to very high Prandtl numbers. Int J Heat Mass Transf 2013;65:907–18. https://doi.org/10.1016/j.ijheatmasstransfer.2013.07.003.

[7] Hamaguchi M, Cardoso M, Vakkilainen E. Alternative technologies for biofuels production in Kraft pulp mills—potential and prospects. Energies 2012;5:2288–309. https://doi.org/10.3390/en5072288.

[8] Ferreira ET de F, Balestieri JAP. Black liquor gasification combined cycle with CO_2 capture – technical and economic analysis. Appl Therm Eng 2015;75:371–83. https://doi.org/10.1016/j.applthermaleng.2014.09.026.

[9] Harvey LDD. Carbon-free energy supply. 2010.

[10] Andersson E, Harvey S. System analysis of hydrogen production from gasified black liquor. Energy 2006;31:3426–34. https://doi.org/10.1016/j.energy.2006.03.015.
[11] Jafri Y, Furusjö E, Kirtania K, Gebart R. Performance of an entrained-flow black liquor gasifier. Energy Fuels 2016;30:3175–85. https://doi.org/10.1021/acs.energyfuels.6b00349.
[12] Cao C, Guo L, Jin H, Cao W, Jia Y, Yao X. System analysis of pulping process coupled with supercritical water gasification of black liquor for combined hydrogen, heat and power production. Energy 2017;132:238–47. https://doi.org/10.1016/J.ENERGY.2017.05.104.
[13] Chiang KY, Lin YX, Lu CH, Chien KL, Lin MH, Wu CC, et al. Gasification of rice straw in an updraft gasifier using water purification sludge containing Fe/Mn as a catalyst. Int J Hydrogen Energy 2013;38:12318–24. https://doi.org/10.1016/j.ijhydene.2013.07.041.
[14] Huet M, Roubaud A, Lachenal D. Conversion of sulfur-free black liquor into fuel gas by supercritical water gasification. Holzforschung 2015;69:751–60. https://doi.org/10.1515/hf-2014-0254.
[15] Cao C, He Y, Chen J, Cao W, Jin H. Evaluation of effect of evaporation on supercritical water gasification of black liquor by energy and exergy analysis. Int J Hydrogen Energy 2018;43:13788–97. https://doi.org/10.1016/J.IJHYDENE.2017.11.158.
[16] Sricharoenchaikul V. Assessment of black liquor gasification in supercritical water. Bioresour Technol 2009;100:638–43. https://doi.org/10.1016/j.biortech.2008.07.011.
[17] Rösch C, Skarka J, Wegerer N. Materials flow modeling of nutrient recycling in biodiesel production from microalgae. Bioresour Technol 2012;107:191–9. https://doi.org/10.1016/j.biortech.2011.12.016.
[18] Wei L, Lu Y. Fluidization behavior in high-pressure water at temperature from ambient to supercritical. Powder Technol 2016;304:89–100. https://doi.org/10.1016/J.POWTEC.2016.08.025.
[19] Loppinet-Serani A, Aymonier C, Cansell F. Current and foreseeable applications of supercritical water for energy and the environment. ChemSusChem 2008;1:486–503. https://doi.org/10.1002/cssc.200700167.
[20] Salimi M, Safari F, Tavasoli A, Shakeri A. Hydrothermal gasification of different agricultural wastes in supercritical water media for hydrogen production: a comparative study. Int J Ind Chem 2016;7:277–85. https://doi.org/10.1007/s40090-016-0091-y.
[21] Fan L. Chemical looping system for fossil energy conversions. 2010. p. 143–214.
[22] Aziz M, Zaini IN, Oda T, Morihara A, Kashiwagi T. Energy conservative brown coal conversion to hydrogen and power based on enhanced process integration: integrated drying, coal direct chemical looping, combined cycle and hydrogenation. Int J Hydrogen Energy 2017;42:2904–13. https://doi.org/10.1016/j.ijhydene.2016.10.060.
[23] Aziz M, Juangsa FB, Kurniawan W, Budiman BA. Clean Co-production of H_2 and power from low rank coal. Energy 2016;116:489–97. https://doi.org/10.1016/j.energy.2016.09.135.
[24] Kumar S. Clean Hydrogen Production Methods 2015;2. https://doi.org/10.1007/978-3-319-14087-2.
[25] Naqvi M, Yan J, Dahlquist E. Synthetic gas production from dry black liquor gasification process using direct causticization with CO_2 capture. Appl Energy 2012;97:49–55. https://doi.org/10.1016/j.apenergy.2011.11.082.
[26] Naqvi M, Yan J, Dahlquist E. System analysis of dry black liquor gasification based synthetic gas production comparing oxygen and air blown gasification systems. Appl Energy 2013;112:1275–82. https://doi.org/10.1016/j.apenergy.2012.11.065.

[27] Darmawan A, Hardi F, Yoshikawa K, Aziz M, Tokimatsu K. Enhanced process integration of black liquor evaporation, gasification, and combined cycle. Appl Energy 2017;204:1035–42. https://doi.org/10.1016/j.apenergy.2017.05.058.
[28] Shen L, Gao Y, Xiao J. Simulation of hydrogen production from biomass gasification in interconnected fluidized beds. Biomass Bioenergy 2008;32:120–7. https://doi.org/10.1016/J.BIOMBIOE.2007.08.002.
[29] Ju F, Chen H, Yang H, Wang X, Zhang S, Liu D. Experimental study of a commercial circulated fluidized bed coal gasifier. Fuel Process Technol 2010;91:818–22. https://doi.org/10.1016/j.fuproc.2009.07.013.
[30] Carlsson P, Wiinikka H, Marklund M, Gr ?, nberg C, Pettersson E, Lidman M, et al. Experimental investigation of an industrial scale black liquor gasifier. 1. the effect of reactor operation parameters on product gas composition. Fuel 2010;89:4025–34. https://doi.org/10.1016/j.fuel.2010.05.003.
[31] Öhrman O, Häggström C, Wiinikka H, Hedlund J, Gebart R. Analysis of trace components in synthesis gas generated by black liquor gasification. Fuel 2012;102:173–9.
[32] Gupta P, Velazquez-Vargas LG, Fan LS. Syngas redox (SGR) process to produce hydrogen from coal derived syngas. Energy and Fuels 2007;21:2900–8. https://doi.org/10.1021/ef060512k.
[33] Goshome K, Yamada T, Miyaoka H, Ichikawa T, Kojima Y. High compressed hydrogen production via direct electrolysis of liquid ammonia. Int J Hydrogen Energy 2016;41:14529–34. https://doi.org/10.1016/J.IJHYDENE.2016.06.137.
[34] Klerke A, Christensen CH, Nørskov JK, Vegge T. Ammonia for hydrogen storage: challenges and opportunities. J Mater Chem 2008;18:2304–10. https://doi.org/10.1039/b720020j.
[35] Vojvodic A, Medford AJ, Studt F, Abild-Pedersen F, Khan TS, Bligaard T, et al. Exploring the limits: a low-pressure, low-temperature Haber–Bosch process. Chem Phys Lett 2014;598:108–12. https://doi.org/10.1016/J.CPLETT.2014.03.003.
[36] Aneke M, Wang M. Process analysis of pressurized oxy-coal power cycle for carbon capture application integrated with liquid air power generation and binary cycle engines. Appl Energy 2015;154:556–66. https://doi.org/10.1016/J.APENERGY.2015.05.030.
[37] Malmali M, Wei Y, McCormick A, Cussler EL. Ammonia synthesis at reduced pressure via reactive separation. Ind Eng Chem Res 2016;55:8922–32. https://doi.org/10.1021/acs.iecr.6b01880.
[38] Cinti G, Frattini D, Jannelli E, Desideri U, Bidini G. Coupling solid oxide electrolyser (SOE) and ammonia production plant. Appl Energy 2017;192:466–76. https://doi.org/10.1016/J.APENERGY.2016.09.026.
[39] Frattini D, Cinti G, Bidini G, Desideri U, Cioffi R, Jannelli E. A system approach in energy evaluation of different renewable energies sources integration in ammonia production plants. Renew Energy 2016;99:472–82. https://doi.org/10.1016/J.RENENE.2016.07.040.
[40] Aziz M, Kansha Y, Tsutsumi A. Self-heat recuperative fluidized bed drying of brown coal. Chem Eng Process Process Intensif 2011;50:944–51. https://doi.org/10.1016/j.cep.2011.07.005.
[41] Hall CW. Handbook of industrial drying. Dry Technol 1988;6:571–3. https://doi.org/10.1080/07373938808916399.

CHAPTER 5

Integrated ammonia production from the empty fruit bunch

Arif Darmawan[1], Muhammad Aziz[2], Muhammad W. Ajiwibowo[3], Muhammad Kunta Biddinika[4], Koji Tokimatsu[5] and Baskoro Lokahita[5]

[1]Agency for the Assessment and Application of Technology (BPPT), Puspiptek Serpong, Tangerang Selatan, Indonesia; [2]Institute of Industrial Science, The University of Tokyo, Meguro-ku, Tokyo, Japan; [3]Department of Mechanical Engineering, University of Indonesia, Depok, Indonesia; [4]Faculty of Industrial Technology, Universitas Ahmad Dahlan, Indonesia; [5]Department of Transdisciplinary Science and Engineering, Tokyo Institute of Technology, Midori-ku, Yokohama, Kanagawa, Japan

Due to population growth and economic development, the world is experiencing a global challenge in managing current and future high energy demand. The extensive overuse of fossil fuels, especially through burning fuel, has resulted in environmental issues such as global warming and air pollution. However, it seems that the only way to leave fossil fuels in the ground is to make a gradual transition. Reduced use of fossil fuels necessitates the development of renewable energy sources that provide a long-term supply with minimal environmental impact. Sustainable energy generation should preferably include three aspects: promotion of renewable energy, reduced environmental risk, and high process optimization. Renewable energy is energy derived from natural resources that can be replenished. Because of their environmental and economic benefits, renewable energy sources such as wind, biomass, solar, geothermal, and hydropower are considered the solution to this issue. Biomass is a sustainable alternative to fossil fuels among the available forms of renewable sources. Biomass, unlike other primary sources, is more uniformly spread across the globe.

In developing and emerging countries, crude palm oil (CPO) is one of the most important commodities, especially across tropical regions such as Southeast Asia, Latin America, and West Africa. It can contribute to the country's GDP and alleviate poverty by providing employment opportunities and increasing the income level. Between 2016 and 2021, the compound annual growth rate of 7.2% in Southeast Asia implies that the market will continue to increase in the future [1]. Because of the massive development of palm plantations and substantial global demand, Indonesia generates the most palm oil globally, more than 30 million tons per year [2].

Innovative Energy Conversion from Biomass Waste
ISBN 978-0-323-85477-1
https://doi.org/10.1016/B978-0-323-85477-1.00006-3

To summarize, CPO processing includes obtaining fresh fruit bunches (FFBs) from plantations, sterilizing and threshing the bunches to extract the palm fruit, mashing the fruit, and pressing out the CPO. These steps result in the production of several palm mill wastes, including empty fruit bunch (EFB), palm kernel shell (PKS), palm kernel cake (PKC), and palm oil mill effluent (POME). Mitigating the negative impacts due to the significant amount of waste disposal is very important since many of these wastes are discarded into the environment. As the main producer countries of CPO, Indonesia and Malaysia have made various efforts to utilize these wastes for value-added goods or energy purposes as well as gaining economic benefits [3,4].

Although PKS and PKC have sparked some interest as potential biomass energy sources, EFB has remained untapped until now. FFB is the main feedstock for CPO production, and EFB accounts for about 23%–25% of it [5], as illustrated in Fig. 5.1. Based on this ratio, approximately 35 Mt of EFB is generated annually around the world. EFB is low density, fibrous, and dense, which presents a major problem for palm milling companies because it takes up a lot of storage space [5]. Many methods and studies have been carried out in order to provide a technological and cost-effective solution to this ever-increasing issue, especially for energy harvesting, such as cocombustion with fossil fuels and thermal treatments. It has been discovered that treating such material will result in value-added products ranging from solid fuels to adsorbents and fertilizers. Furthermore, these fuels are said to be able to reduce a country's industrial and commercial energy costs [6]. Like palm oil mills' press fiber and shell, EFB has traditionally been used as a solid fuel for steam boilers. The steam produced is used to power turbines that generate electricity. However, the "white smoke" issue caused by the EFB's high moisture content (>60%) has an aesthetic effect on the beautiful surroundings [7].

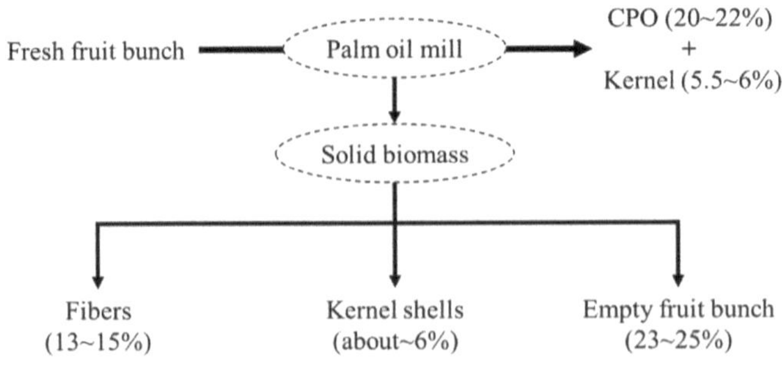

Figure 5.1 Crude palm oil processing from the fresh fruit bunch.

EFB from palm oil processing can also be transformed into biofuels using thermochemical or biological conversion methods. Both routes of conversion involve pretreatment of EFB, with alkali pretreatment being the most efficient. Fig. 5.2 illustrates a review of EFB conversion technology. Gasification appears to be the most appropriate method for extracting bioenergy from EFB and has the potential for commercialization, according to studies on biomass thermochemical conversion. Pyrolysis, alternatively, yielded a very complex bio-oil with high viscosity and water content, making commercialization difficult. On the other hand, charcoal made from EFB pyrolysis has the potential to be a commercial product. Overall, considering its high water content and low energy capacity, EFB may not be an excellent candidate for solid fuel production compared to other palm oil residues such as oil palm kernels, even after the torrefaction process [8]. Another way to get biofuels from EFB is through biological conversion. The most intensive research was done on cellulosic ethanol production, which yielded about 50 g/L titers with a 20% (w/v) biomass loading via NaOH pretreatment. The ammonia fiber expansion (AFEX) also showed promise in EFB pretreatment, with a glucose yield of 90% achieved with a biomass loading of 9%. The water extract of the AFEX pretreated EFB was highly fermentable. EFB also showed some encouraging preliminary results in the development of ABE (acetone, butanol, and ethanol) and biogas, but further research is required to boost EFB conversion potential in these areas.

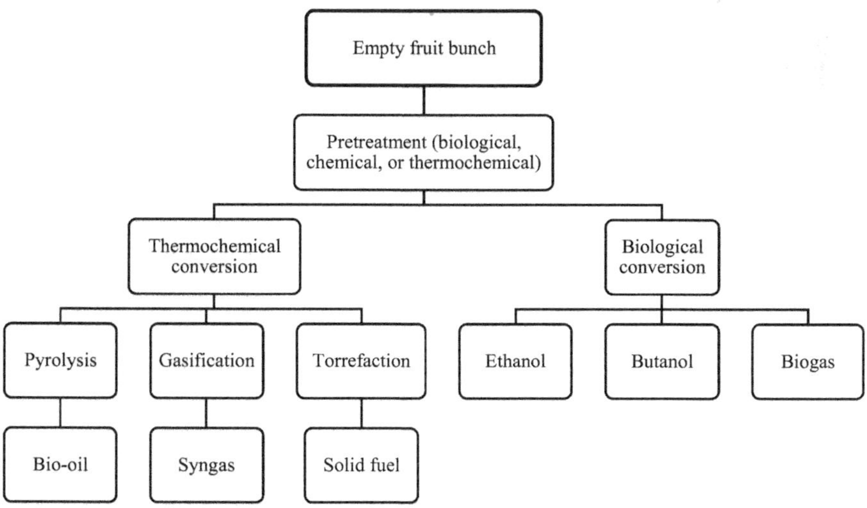

Figure 5.2 Optional routes for empty fruit bunch (EFB) conversions into biofuels [8].

Biomass feedstock is not suitable for direct use by end-users as a source of primary energy. They can use various conversion technologies to produce secondary energy forms or energy carriers such as biofuels, heat, or electricity. This situation has sparked a lot of interest in secondary energy carriers like hydrogen (H_2), potentially helping renewable energy in different forms. Many researchers have been drawn to the production of H_2 in order to effectively manufacture, store, transport/distribute to end-users, and use it with minimal environmental impact at the point of use. H_2 can be used for both short and long periods if provided by sufficient storage and proven infrastructures.

Hydrogen (H_2) is an emerging energy carrier (secondary energy source), including biofuels derived from EFB. From syngas, which consists mainly of hydrogen and carbon monoxide, the water–gas shift reaction can then be applied to transform the generated CO into H_2 and CO_2. As compared to other secondary sources of energy, hydrogen has many advantages as a fuel, including a high gravimetric energy content of 120 MJ/kg, a high reactivity rate, and a very high flame speed during combustion. Furthermore, hydrogen's value chain is rapidly evolving in developed countries such as Japan [9]. Fuel cells, direct combustion, and mixing with other fuels are the most common hydrogen conversions at the final users. The world is increasingly evolving into a hydrogen society. However, hydrogen has a range of disadvantages and challenges, including storage safety, a low volumetric energy density, and transportation.

Conventionally, hydrogen is stored mainly by compressing it to extremely high pressures. Hydrogen cannot be stored at atmospheric pressure due to its small molecule size, which allows it to easily slip through storage walls and cause material embrittlement [10]. This circumstance puts the storage capability's safety in jeopardy. As a result, several attempts have been made to improve hydrogen's safety by various storage methods. One of the most promising approaches is to store hydrogen by converting it to other stable chemicals such as Ammonia (NH_3) and methylcyclohexane (MCH).

5.1 Ammonia for hydrogen storage

Because of its high performance, high hydrogen density, applicability, and ease of catalytic decomposition, ammonia is a very appealing hydrogen storage system [11,12]. It is also thought to have a high economic output due to its well-developed synthesis and distribution technology.

Unfortunately, due to its low volumetric density (only around 3 Wh/L at atmospheric conditions) and prone to leakage, hydrogen poses a major storage challenge. To address this issue, efficient hydrogen storage and transportation methods are being actively developed, including compression, liquefaction, hydrate binding, and conversion to other materials (ammonia, methanol.). Technology maturity, economic viability, and the readiness of the supporting technology should all be considerations in the storage. Based on the aforementioned aspects, liquid hydrogen, MCH, and ammonia are considered potential hydrogen storages among several alternatives.

Ammonia has the highest hydrogen storage density (as much as 17.8 wt%t) than other hydrogen storage media. Furthermore, relative to liquid hydrogen and MCH, ammonia has the largest volumetric hydrogen density. Unfortunately, ammonia is poisonous and corrosive, necessitating careful storage and transportation. However, since ammonia can be easily detected, even at concentrations as low as one part per million, the leakage of ammonia can be monitored, decreasing the risk of toxicity. Ammonia storage has several advantages, including the ability to use it directly, a low-cost energy carrier (potentially), and existing regulations and infrastructure. Ammonia storage, however, has lower reactivity than hydrocarbons, necessitates treatment due to toxicity and pungent odor, necessitates treatment and management by certified engineers, and consumes a significant amount of energy during dehydrogenation (roughly 13% of hydrogen energy) and purification. Ammonia can be used directly in combustion and fuel cells or decomposed first to hydrogen and nitrogen. The hydrogen can then be used in a wider range of applications, including hydrogen fuel cells (Table 5.1).

Direct use of ammonia is considered beneficial in terms of energy efficiency and economic performance because decomposition is also very energy-intensive. The Haber–Bosch process (HBP), electrochemical process, and membrane reactors are the most common methods for generating ammonia. Electrochemical processing is expected to use 20% less energy than the HBP. While electrochemical processing has the benefit of requiring less energy per generated NH_3, the technology is still very costly to implement. On the other hand, due to the high energy needed to dissociate the triple-bonded nitrogen molecule, the HBP requires a high temperature (400–600°C) and a high pressure (up to 30 MPa). The HBP is considered the most mature and technologically wise technology. The possible schematic route for ammonia synthesis via the HBP is shown in Fig. 5.3.

Table 5.1 Comparison among potential hydrogen storages (liquid hydrogen, methylcyclohexane (MCH), and ammonia).

Characteristics	Liquid H_2	Toluene-MCH	Ammonia
Properties			
Molecular weight	2.016	98.19	7.03
Density (kg/m)	70.8	769	682 (0.1 MPa)
Boiling point (°C)	−252.9	101	−33.34
Gravimetric H_2 density (wt%)	100	6.16	17.8
Volumetric H_2 density (kg-H_2 m^{-3})	70.9	47.1	120.3
Hydrogen release temperature (°C)	−252.9	200−400	350−900
Regeneration temperature (°C)	—	100−200	400−600
Ignition temperature (°C)	571	535 (tol), 283 (MCH)	651
Enthalpy change in H_2 release (kJ/mol)	0.899	67.5	30.6
Physical characteristics	High storage capacity (800 times, based on volumetric density)	High storage capacity (500 times, based on volumetric density)	Very high storage capacity under the pressure of about 800 kPa or temperature of −33°C (1200 times, based on volumetric density)
Infrastructure	Requires technological development and construction for large scale	Possible utilization of existing gasoline infrastructure	Possible utilization of existing propane infrastructure
Utilization	- Hydrogen combustion - Fuel cell	- Hydrogen combustion (after being dehydrogenated) - Fuel cell (after being dehydrogenated and purified)	- Direct combustion - Fuel cell (after being dehydrogenated and purified) - Direct ammonia fuel cell

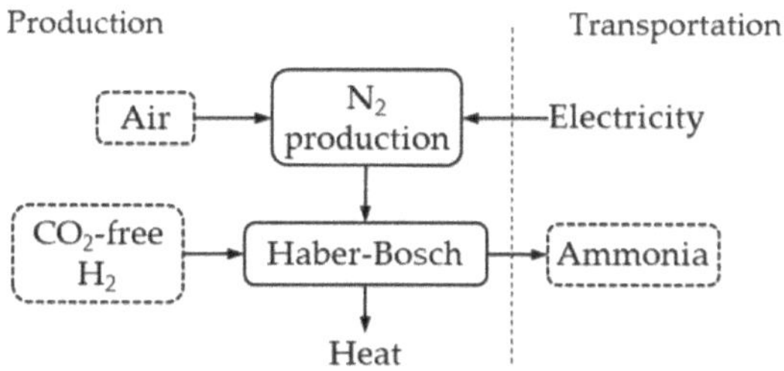

Figure 5.3 Schematic route of ammonia production via Haber–Bosch process.

Fig. 5.4 depicts the economic cost forecast for each hydrogen storage system; liquid hydrogen, MCH, ammonia (direct and indirect use). Further assessment is carried out based on analysis by Mizuno et al. [13], taking into account some additional aspects. The decomposition of ammonia to hydrogen can be avoided in the case of ammonia for direct use. As a result, the additional cost can be eliminated as well. Hydrogen is expected to be produced in Australia and exported to Japan due to the current energy market and the bilateral relationship. The assumptions used during cost calculation and forecast are listed in Table 5.2. Since the decomposition of NH_3 to H_2 is unnecessary in the case of NH_3 for direct use, the decomposition cost can be reduced, resulting in a lower cost than NH_3 with decomposition. The estimated CIF (cost, insurance, and freight) cost of hydrogen in 2030 is 30 JPY.Nm^3, according to the NEDO roadmap, and can be further lowered to 20 JPY.Nm^3 in the future. This target appears to be feasible if NH_3 for direct use is preferred as the primary hydrogen storage, as the total costs in 2030 and 2050 are approximately 24 and 21 JPYNm^3-H_2, respectively [14].

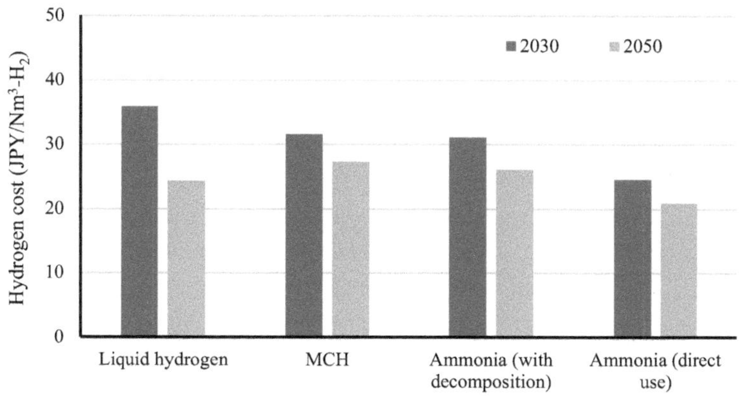

Figure 5.4 Cost forecast of each hydrogen storage [14].

Table 5.2 Main assumptions of economic calculations.

	Value	Note/Ref.
General conditions		
Hydrogen supply amount (t/y)	3×10^5	[15]
Hydrogen price (JPY/Nm)	10 (2030) and 8 (2050)	[15]
Transport distance (km)	9000	Australia to Japan
Equipment utilization rate (%)	90	
Equipment operation period (y)	30	
Average electricity price (JPY/kWh)	127	Australian price
Liquefaction		
Shipping fuel	Boil-off H_2	[16]
Tanker volume (m^3)	160,000	
Boil-off rate (% d^{-1})	0.2	
Evaporation rate at loading (%)	1.3	
Evaporation rate at discharge (%)	1.2	
Specific energy consumption (kWh kg-H_2)	15	
Toluene hydrogenation		
MCH production (t/y)	5.02×10^6	Avg. price in 2017 45 JPY/L [17]
Shipping fuel	Diesel MFO	
Ammonia synthesis		
Nitrogen production (t/y)	1.39×10^6	
Energy for nitrogen production (kWh Nm^3-N_2)	0.38	
Ammonia production (Nm^3/d)	1.69×10^6	

5.2 Studies on ammonia production

Aziz et al. [18] explored the possibility of combining chemical looping gasification and the HBP for NH_3 production from low-rank coal (Fig. 5.5). Coal direct chemical looping, nitrogen extraction, ammonia synthesis, and power generation are key elements of the integrated process. In order to remove the amount of water in the low-rank coal, it is first evaporated. Three circulated reactors are used in coal direct chemical looping: reducer, oxidizer, and combustor. The pulverized LRC has generated CO_2 by reacting directly with the oxygen carrier in the reducer. In an oxidizer, the reduced oxygen carrier is also reacted with steam to generate hydrogen. Finally, the oxygen carrier is oxidized in the combustor

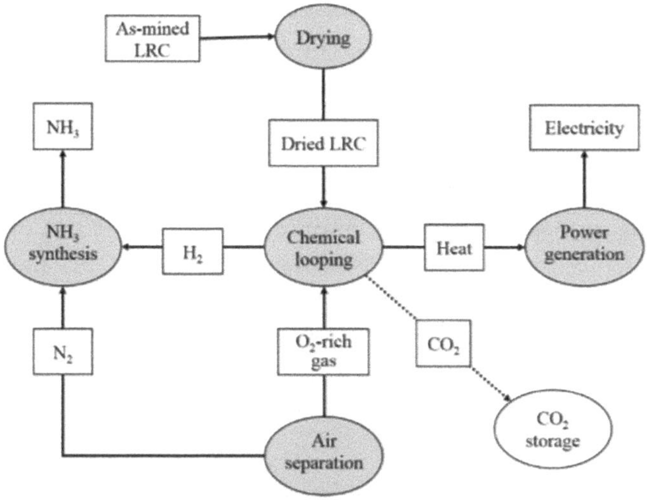

Figure 5.5 The development of an integrated system for NH_3 production from low-rank coal.

with O_2-rich gas to return to its original state. In the NH_3 synthesis, the generated hydrogen is reacted with N_2 produced in the N_2 separation unit. Their proposed system had a 75% energy efficiency, but they did not explore the possibility of using a renewable energy source. Also, Reese et al. [19] carried out an experimental study of an integrated NH_3 production system that used excess electricity from renewable sources, especially wind power. In an experiment with the end-to-end method, they validated their design. They did not, however, look into the possibilities of process intensification and heat integration any further.

Andersson and Lundgren also performed a technoeconomic evaluation of integrated NH_3 production in a pulp and paper mill using biomass gasification [20]. The integrated NH_3 production system involved biomass gasification, a water-gas shift, and ammonia synthesis. The gasification model was created in Aspen Plus and was based on a pressurized entrained flow biomass gasifier technology. The hydrogen concentration of generated syngas was increased using the water-gas shift (WGS) reactor. The NH_3 synthesis was carried out over an iron-promoted catalyst, which was modeled using a Gibbs reactor with a 440°C outgoing gas temperature. The conversion efficiency is typically 20%–30% per pass, and unreacted gases must be recycled to increase the NH_3 production yield. They also analysed the economic implications of such systems. However, no significant effort has yet been made to design and study the possibility of process integration, particularly in NH_3 production from biomass-based hydrogen. Arora et al. [21] used life cycle costing and life cycle assessment approaches to determine the enviro-economic efficiency of NH_3 production from biomass in various countries. They did, however, just look at the new method. There is no attempt to model and evaluate the advanced method, which increases energy efficiency and decreases greenhouse gas emissions significantly. Gilbert et al. [22] also examined the effect of a biomass-based NH_3 production method on greenhouse gas reduction. When the gasification method is used instead of steam reforming, they discovered a massive opportunity for greenhouse gas reductions of up to 65%. While replacing gasification with steam reforming is a promising way to reduce greenhouse gas emissions, the evaluated system still has many improvement opportunities.

5.3 Efficient ammonia production from empty fruit bunch via hydrothermal gasification, syngas chemical looping, and NH_3 synthesis

This chapter describes and analyzes an integrated system to produce NH_3 from EFB with a strong motivation to respond to the above perspectives and significantly improve biomass-based NH_3 production, particularly in terms of system performance. Supercritical water gasification (SCWG), syngas chemical looping, and NH_3 synthesis are three of the major technologies that are simultaneously integrated. Exergy recovery and process integration are also used during analysis and simulation to enhance efficient heat circulation in the integrated system. The next section goes into the particular methods used in each subsystem. The proposed system's performance is defined and discussed in the results and discussion section, particularly in terms of each system's system efficiency and the integrated system overall. In order to obtain the system's optimized operating parameters, certain thermodynamic parameters are modified.

The system's overall configuration is represented in Fig. 5.6. To reduce EFB and POME transportation costs, the design is set to be directly attached to an existing palm oil processing facility. The processed fruit bunch is directly fed into the integrated NH_3 production system after processing, along with POME and makeup water, by entering the SCWG process. Following that, H_2-rich syngas is generated. The syngas is then transformed to pure hydrogen, then fed directly into the NH_3 synthesis system. Power and heat are then recirculated to intensify each process and generate excess power from the combined processes.

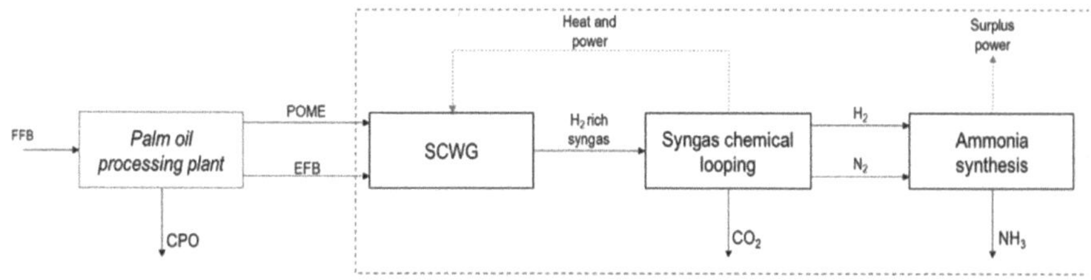

Figure 5.6 Conceptual design of the integrated palm oil H_2 and NH_3 production.

5.3.1 The general assumption for the calculation

For thermodynamic, energy, and mass balance calculations, a steady-state Aspen Plus V8.8 process simulator is used. The parameters for each subsystem were extracted from other researchers' experimental studies. Some assumptions are taken into account in this analysis, such as: the ambient temperature being 30°C, no heat loss anticipated, no leaks or pollution expected. In addition, the ambient air is set to be 79% N_2 and 21% O_2. Furthermore, the EFB input is estimated to be between 8 and 40 t/h, based on Indonesia's daily CPO output of 94,500 t/d. The overall process flow diagram of the system is depicted in Fig. 5.7. The main heat

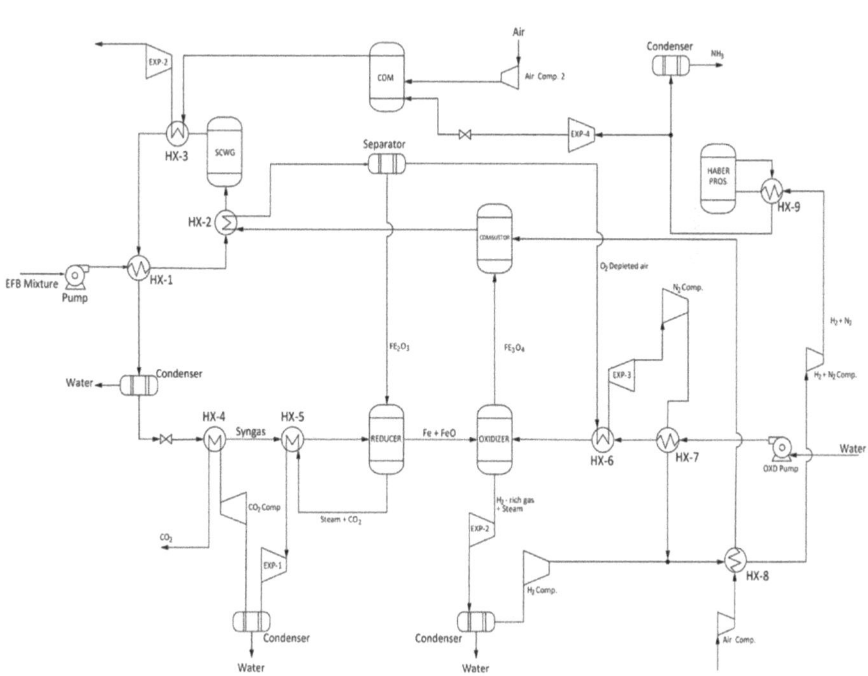

Figure 5.7 Overall system process flow diagram of NH_3 production from empty fruit bunch (EFB).

exchanger that is evaluated is the superheater, which is used in the SCWG process. The SCL method allows the use of the other heat exchanger systems.

5.3.2 Supercritical water gasification of empty fruit bunch for syngas production

It is important to establish a thermochemical route to transform biomass, or in this case, EFB, into a high-quality gaseous substance. SCWG is used as the primary conversion technology throughout this analysis to convert EFB to syngas. SCWG is characterized by a high yield of hydrogen during the reaction, as observed in other studies. Sivasangar et al. [23] analyzed an experimental SCWG in which POME was used as the reaction medium. POME is a residue that is normally discarded after processing because it contributes little value to the industry. Their research found that using POME for this process is feasible and yielded roughly the same hydrogen concentration as using water. Furthermore, adopting the SCWG method has the additional benefit of not needing any predrying of wet feedstock EFB.

In addition to POME, EFB and makeup water are pumped directly into the gasifier. At about 380°C and 150 bar, a superheater is used to raise the mixture to a supercritical state. After that, the gasifier generates H_2-rich syngas. This H_2-rich mixture is then fed into the SCL process, which is discussed in more detail in the following section. The adopted SCWG method is represented in Fig. 5.4 as a conceptual view. Simultaneous hydrolysis (Reaction 5.1), steam reforming (Reactions (5.2) and (5.3)), water-gas change (Reaction (5.4)), and methanation (Reaction (5.5)) are the main reactions in the SCWG process. The feedstock's chemical energy is converted into a syngas mixture containing mainly H_2, supplemented by CO_2, CH_4, and CO (Fig. 5.8).

$$\left[C_XH_{y-2}O_{Z-1}\right]_n + {}_nH_2O \rightarrow {}_nC_XH_YO_Z \tag{5.1}$$

$$C_XH_yO_z + (2x - z)H_2O \rightarrow {}_xCO_2 + \left(2x - z + \frac{y}{2}\right)H_2 \tag{5.2}$$

$$C_XH_yO_z + (x\text{-}z)H_2O \rightarrow {}_xCO + \left(x - z + \frac{y}{2}\right)H_2 \tag{5.3}$$

$$CO + H_2O \rightarrow CO_2 + H_2 \tag{5.4}$$

$$CO + 3H_2 \rightarrow CH_4 + H_2O \tag{5.5}$$

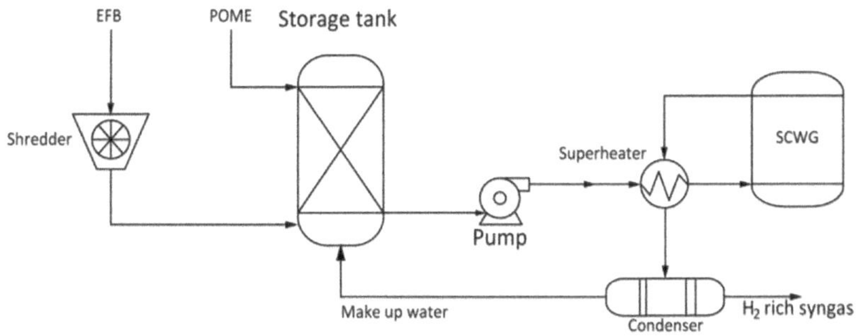

Figure 5.8 Conceptual view of the supercritical water gasification (SCWG) process.

The RGibbs-type reactor and a heat exchanger are used in the ASPEN Plus process simulator to model the gasification and superheating processes. The simulation's main objective is to determine how much heating is needed to help the SCWG operation. The syngas production is taken from Sivasangar's experimental research [23] (Table 5.3).

5.3.3 Syngas chemical looping

Metal oxides are used as a reaction agent in syngas chemical looping (SCL), which is a multireactor process. It is a development of the chemical looping combustion (CLC) technology, which provides a clean way to oxidize fuel while allowing in situ CO_2 capture and storage at low to no energy cost. The SCL process uses three reactors to produce H_2, including the oxidizer and reducer, and a third reactor called the combustor. The process begins in the reducer, where metal oxide oxidizes the syngas' hydrocarbon content. Steam is then fed into the reactor from the oxidizer, where it reacts with the remaining syngas to produce a high concentration of H_2-rich gas. In the

Table 5.3 Elemental composition of empty fruit bunch (EFB) and palm oil mill effluent (POME) used in the study.

Properties	EFB (wt%) [24]	POME (mg/L) [25]
Elements	C: 50.2; H: 7.1; O: 41.87; N: 0.82; S: 0.01; ash: 4.8	Oil and grease: 4000–5000 Total solids: 40,500 Suspended solids: 18,000 Total volatile solids: 34,000 Total nitrogen: 750 P: 180; K: 2270; ca: 439; Bo: 7.6; Fe: 46.5; Mn: 2; Cu: 0.89; Mg: 615; Zn: 2.3

Table 5.4 Syngas chemical looping (SCL) process modeling parameters.

Process	Reactor type	Calculation model	Operating condition
Reduction	Counter-current moving bed	RStoic	750–900°C; 25–35 bar
Oxidation	Counter-current moving bed	RStoic	500–750°C; 25–35 bar
Combustion	Circulating fluidized bed	RStoic	900°C; 25–35 bar

combustor, the reduced metal oxide is also reoxidized to its original state. A mixture of iron oxide (Fe_2O_3) and heat diluents consisting of aluminum oxide (Al_2O_3) and silica carbide (SiC) at 70% and 30%, respectively, is assumed in this analysis. Table 5.4 defines the reactor type and the design considerations that go with it in the Aspen Plus software.

The SCL process's simple description is shown in Fig. 5.9 as described in Table 5.4; both the reducer and the oxidizer adopted a counter-current moving bed reactor scheme. The combustor, on the other hand, adopted an entrained bed reactor scheme. These reactor types are chosen based on the currently operating pilot-scale SCL plant [26].

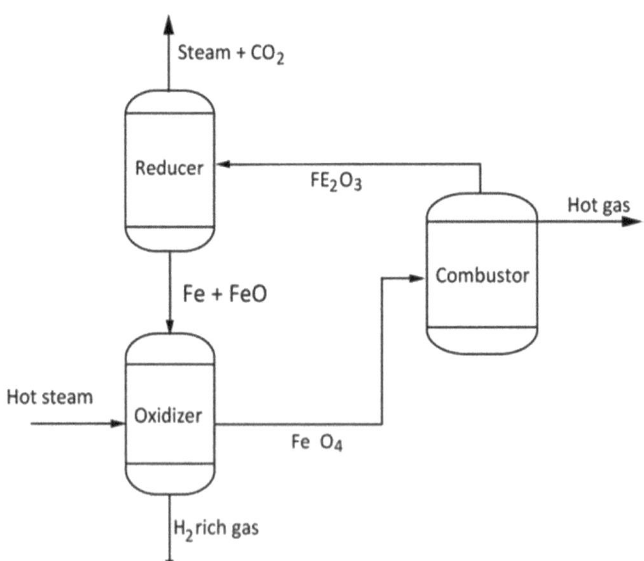

Figure 5.9 Simple flow diagram of the syngas chemical looping (SCL) process.

The H_2-rich syngas is fed into the reducer after being cleaned, where it is reacted with a metal oxide mixture of iron and other heat diluents. It is normally operated at a temperature of 750–900°C and a pressure of 25–35 bar. CO_2 and steam are formed entirely from the syngas mixture. The metal oxide mixture is then reduced down to Fe and FeO. Reactions (5.6) to (5.10) identify the reactions involved.

$$Fe_2O_3 + CO \rightarrow 2FeO + CO_2 \tag{5.6}$$

$$FeO + CO \rightarrow Fe + CO_2 \tag{5.7}$$

$$Fe_2O_3 + H_2 \rightarrow 2FeO + H_2O \tag{5.8}$$

$$FeO + H_2 \rightarrow Fe + H_2O \tag{5.9}$$

$$4Fe_2O_3 + 3CH_4 \rightarrow 8Fe + 3CO_2 + 6H_2O \tag{5.10}$$

The syngas composition, reaction temperature, and overall reduction rate influence the amount of heat produced in the reducer. As a consequence, these parameters affect future process heat integration. The heat generated in this reactor will be directly affected by whether the reaction is endothermic or exothermic. In addition, the metal oxides completely oxidize the syngas mixture, leaving only CO_2 and steam in the exhaust gas. Thus, removing the exhaust gas's heat makes it possible to condense the steam and leave only CO_2. As a result, CO_2 generated in this process can further be compressed, stored, and distributed directly at a low energy cost. The metal oxide mixture is fed into the oxidizer after being reduced. The steam is reacted with metal oxides to generate H_2 and a pure H_2 mixture is extracted by steam condensation. The following are some of the reactions:

$$Fe_2 + H_2O \rightarrow FeO + H_2 \tag{5.11}$$

$$3FeO + H_2O \rightarrow Fe_3O_4 + H_2 \tag{5.12}$$

The steam for this process comes from the cooling process of the previous reactor, namely the reducer. The oxidizer works at the same pressure and temperature as the reducer, which is 25–35 bar and 500–750°C. In addition, the process continues in the combustor, where the Fe_3O_4 is further reduced and regenerated back to its original Fe_2O_3 state. The metal oxide mixtures and pressurized air are the two main streams that are pumped into this reactor. This process generated a lot of heat and is an excellent addition to the process integration scheme. The following are the common combustion reactions:

$$4Fe_3O_4 + O_2 \rightarrow 6Fe_2O_3 \quad (5.13)$$

The extracted air is used for electricity generation through the expanders, which also acts as a heat diluent and has a high temperature and pressure. The provided heat and electricity are used to offset the need for auxiliary power and heat. Based on the LHV approach, the generated H_2 is compared to the input syngas to determine the SCL process's efficiency. Eq. (5.14) explains how to perform the calculation. In Eq. (5.15), the power generated and consumed can also be compared. Aside from the different EFB to water ratios, the various operating pressures are also examined.

$$\eta_{SCL} = \frac{\dot{m}_{H_2} \times LHV_{H_2}}{\dot{m}_{syngas} \times LHV_{syngas}} \quad (5.14)$$

$$\eta_{SCWG-SCL} = \frac{\left(\dot{m}_{H_2} \times LHV_{H_2}\right) - \text{Heating requirement}}{\dot{m}_{EFB} \times LHV_{EFB}} \quad (5.15)$$

$$W_P = W_{EXP1} + W_{EXP2} + W_{EXP3} + W_{EXP4} + W_{EXP5} \quad (5.16)$$

$$W_C = W_{OXD-Pump} + W_{EFB-Pump} + W_{Air\ Comp\ 1} + W_{Air\ Comp\ 2} + W_{H_2\ Comp} + W_{N_2\ Comp} + W_{H_2+N_2\ Comp} + W_{CO_2} \quad (5.17)$$

5.3.4 Haber process for NH_3 production

The HBP, also known as the Haber process, is one of the most popular ways to produce NH_3, as outlined in the previous part. The process needs a temperature of 400°C and a pressure of about 150 bar to perform the reaction properly. The reaction generates heat in the process, which can be utilized if the process is properly integrated. The reversible reaction is as follows:

$$H_2 + N_2 \rightarrow NH_3 \quad (5.18)$$

The basic process flow of the formulated HBP is shown in Fig. 5.10. The RStoic reactor model is used to model this mechanism in ASPEN Plus. N_2 and H_2 generated during the SCL process are among the input streams. With stream recycling to provide heat integration for the prior processes, the conversion ratio is assumed to be 30% per cycle.

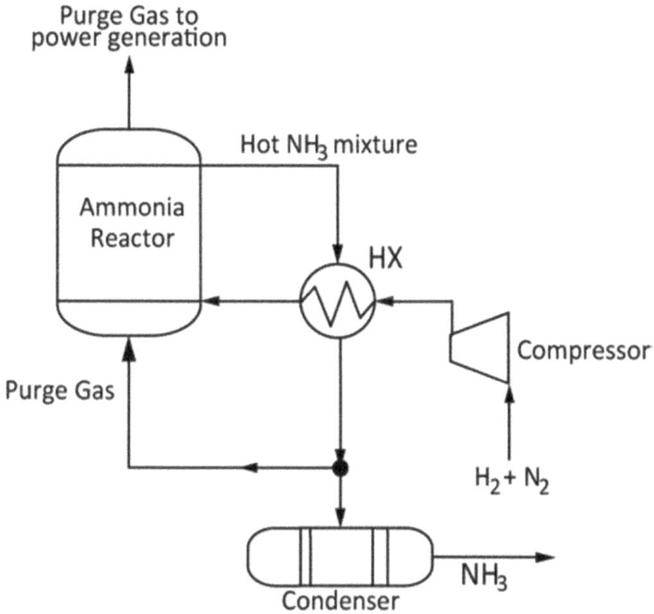

Figure 5.10 Basic flow diagram of the Haber–Bosch process.

Recycled gas and purge gas are supplied from the unreacted gas that still contains H_2 and N_2. The former is used to purge the reactor, while the latter is recirculated into the NH_3 processing reactor. The purge gas is used for combustion after the purging process, primarily for heat and power generation. Ninety percent of the unreacted gas is used as a purge gas in this system. This amount is required so that enough heat can be produced to support previous processes, especially the EFB SCWG, which uses a lot of water.

Eq. (5.19) is also used to evaluate the performance of the NH_3 production process. It compares the HHV of generated NH_3 with the HHV of raw EFB. It can also be referred to as total efficiency because it measures the process's cumulative output and input. In addition, Eq. (5.20) compares the same thing to obtain the total energy efficiency, but it also takes into account the power generated and consumed in the overall system.

$$\eta_{NH_3} = \frac{\dot{m}_{NH_3} \times HHV_{NH_3}}{\dot{m}_{EFB} \times HHV_{EFB}} \tag{5.19}$$

$$\eta_{H_2-overall} = \frac{\left(\dot{m}_{H_2} \times HHV_{H_2}\right) + (W_P - W_C) - \text{Heating requirement}}{\dot{m}_{EFB} \times HHV_{EFB}} \tag{5.20}$$

5.3.5 System analyses

Each system's efficiency is assessed based on the operating parameters to provide a technical view of the integrated system's performance, as described in the previous parts. To begin, Fig. 5.11 depicts the effect of the EFB-to-water ratio on H_2 production quality. The SCL's obtained efficiency is typically in the 70%–77%. The efficiency is calculated by comparing the heating values of the emitted H_2 to the input syngas' heating values. With an EFB-to-water ratio of 2.5%, the highest performance is estimated to be about 76.3%. At an EFB-to-water ratio of 6.25%, the lowest is 71.9%. The steam iron process in the oxidizer is the primary mechanism for H_2 formation. The amount of H_2 emitted by steam reaction increases as Fe and FeO's amount inserted into the reactor increases.

In the case of the 6.25% EFB-to-water ratio, it is supposed that the syngas had a higher CO content at the start than the other EFB-to-water ratios. Through the reducer reduction phase, a higher syngas content will produce more Fe and FeO. As a result, in the downstream oxidizing phase,

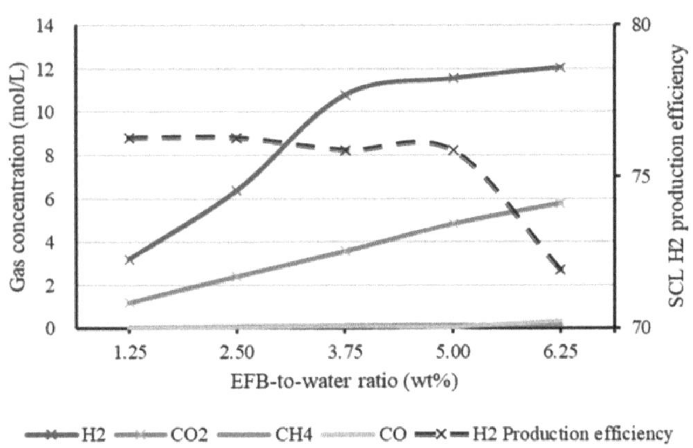

Figure 5.11 Effect of empty fruit bunch (EFB) to water ratio to the syngas composition and the syngas chemical looping (SCL) H_2 production efficiency.

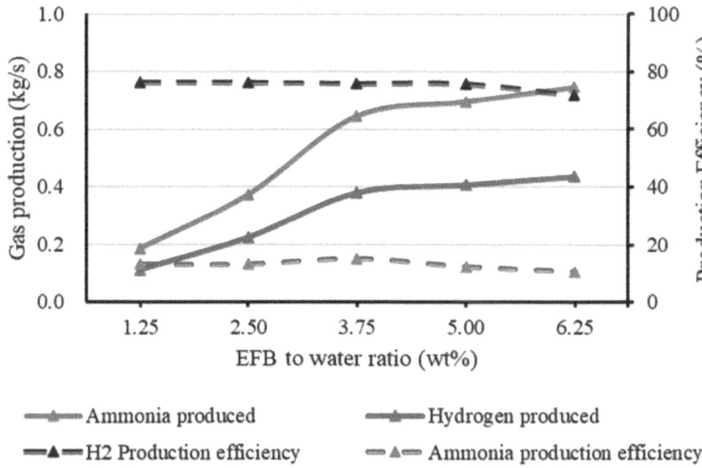

Figure 5.12 Effect of empty fruit bunch (EFB) to water ratio to the H_2 and NH_3 production and their efficiencies.

more H_2 can be generated. Although the syngas in the other ratios contains substantial amounts of H_2 and CO_2, this does not contribute significantly to Fe and FeO's yield in the steam iron process.

In addition, Fig. 5.12 shows a comparison of H_2 and NH_3 output as well as their respective efficiencies for each EFB-to-water ratio. The SCL efficiency is described as the H_2 output efficiency. The NH_3 production efficiency, on the other hand, compares the heating values of the generated NH_3 to the raw EFB. At an EFB-to-water ratio of 3.75%, the highest output rate for both H_2 in the SCL and NH_3 is achieved. The quantity of H_2 in the syngas itself is the primary cause of this phenomenon. Purer H_2 can be obtained from the SCL with a larger amount of H_2 in the syngas. Thus a larger amount of NH_3 can be generated. When it comes to NH_3 output, a higher EFB input reduces efficiency significantly. It is primarily due to NH_3's low heating value. At a 2.50% ratio, the highest NH_3 output efficiency is achieved. At a ratio of 3.75%, the highest output efficiency for both H_2 and NH_3 is achieved.

In Fig. 5.13, the overall generated and consumed power in the system is compared for each EFB-to-water ratio. Overall, extra power is required to support the system for each EFB-to-water ratio because the consumed power is higher than the power generation, which is around 6−11 MW on average. Purge gas recovery generates the most electricity. The H_2 and N_2 compression, on the other hand, consumes the most power, leaving the

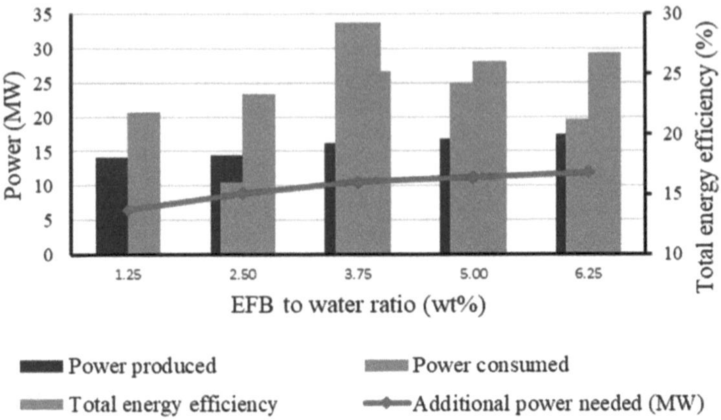

Figure 5.13 Produced power, consumed power, the additional power required in the system, and the total energy efficiency for each EFB-to-water ratio.

SCL to the HBP, which demands enormous power due to the high pressure required for the reaction. However, in the ratio of the produced H_2 to the required input heating, additional power, and the input EFB's HHV, the total achieved energy efficiency is relatively high, reaching a maximum efficiency of 29.3%. This phenomenon has been observed in other SCWG studies. The cold gas efficiency (CGE) can vary depending on the situation. The initial CGE is achieved at 40%–50% in this case. As a result, the total energy efficiency obtained will not exceed this CGE. On the other hand, other researchers believe it is possible to achieve a produced gas ratio greater than 100% in terms of mass gain and energy efficiency.

Changing the system's pressure produced similar results, with the consumed power being much larger than the produced power. Although, since the system generates slightly more power, increasing the operating pressure will slightly improve overall system efficiency. The effect of varying the operating pressure on the produced and consumed power throughout the system is shown in Fig. 5.14. Incorporating a more efficient gasification system is one of the main ways to improve the system's efficiency.

Finally, the system will achieve a maximum EFB to NH_3 efficiency of over 15% and a 76.2% H_2 conversion efficiency in the SCL module and a 46.3% overall syngas to H_2 conversion efficiency. It is obvious to see how the SCWG system uses a lot of heat. Furthermore, if the separated CO_2 is collected, the process will achieve 100% CO_2 capture and storage pressure at 200 bar. If CO_2 is not stored, the separated CO_2 with high pressure can

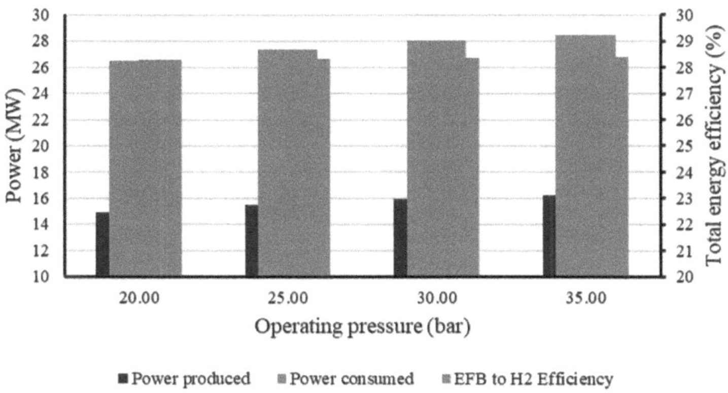

Figure 5.14 Effect of alternating operating pressure on the produced and consumed power in the system.

be recovered for other electricity production by extracting it in an expander, resulting in a 4% increase in energy efficiency. In this report, the SCWG's CGE is relatively low, indicating how high the heating demand is, as well as indicating one of the crucial requirements in designing an integrated SCWG system. Due to the highly efficient SCL method, this device can produce H_2 at a far higher efficiency than other EFB recovery systems. Nonetheless, in order to increase overall performance, an efficient syngas production method must be developed. Furthermore, it is demonstrated that directly converting the generated H_2 into NH_3 is feasible and has numerous potentials for improving energy efficiency. This project was based on empirical results and could serve as a basis for future work involving the integration of SCWG and SCL to generate H_2 and/or NH_3.

5.4 Direct ammonia production via a combination of carbonization and thermochemical cycle from the empty fruit bunch

As previously mentioned, H_2 poses additional storage and transportation challenges due to its low volumetric density. Both technology and the economy need an effective generation and storage system. A few H_2 storages have been assessed and evaluated to determine the most effective way to increase the fuel's volumetric energy density. Three potential methods to store H_2 are reported, including liquid H_2, MCH, and ammonia, especially in terms of their physical characteristics, technological feasibility for application, and projected cost [14]. Due to its high performance, high density (4 kWh/kg), excellent applicability, such as for direct fuel cells, and ease of decomposition, ammonia (NH_3) is considered a quite promising method of storing H_2 among several alternatives. In comparison to other storage methods, NH_3 has a higher ignition temperature, which means it is safer to store.

Sajid and Bicer [27] introduced an integrated system for more environmentally friendly NH_3 synthesis that uses H_2 generated by solar thermal methane cracking. A solar thermal cracking (STC), chemical looping combustion (CLC), NH_3 synthesis, and a single-stage absorption cooling system are all part of the proposed system (SSAC). Solar energy is becoming an essential component of producing renewable hydrogen and reducing carbon emissions. STC is the thermal decomposition of methane or other hydrocarbon fuels into hydrogen and carbon (illustrated in Fig. 5.15). CLC is a method of trapping CO_2 without sacrificing the plant's net production. The CLC is made up of two reactors: a fuel reactor and an air reactor. They reported that in the system, solar thermal cracking has the highest exergy destruction rate of 20.9 MW, with overall device energy and exergy efficiencies of 28.8% and 25.8%, respectively.

The utilization of solar energy for NH_3 synthesis has also been studied by Siddiqui and Dincer [28]. The schematic of the proposed integrated ammonia synthesis and fuel cell system for solar power plants is illustrated in Fig. 5.16. In the system, excess power from a solar photovoltaic system is used to generate H_2 through polymer electrolyte membrane (PEM) electrolysis. Furthermore, the nitrogen gas is generated by a pressure swing adsorption (PSA) based air separation system. Furthermore, the oxygen generated in the PSA subsystem is stored for use in the direct form ammonia fuel cell (AFC) later on. Then, the ammonia synthesis reactor receives the produced and stored gases (hydrogen and nitrogen). However, solar energy can be highly variable due to weather and geographical location, which is very well known. They found that overall energy efficiency ranged from 15.68% to 15.83% during the year measured.

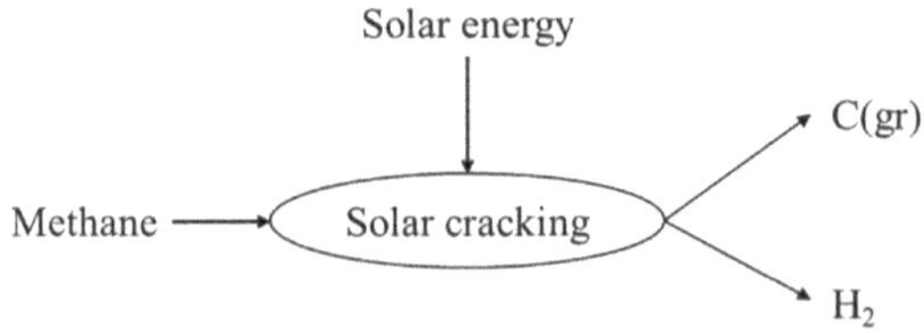

Figure 5.15 Principle of solar thermal cracking adopted [27].

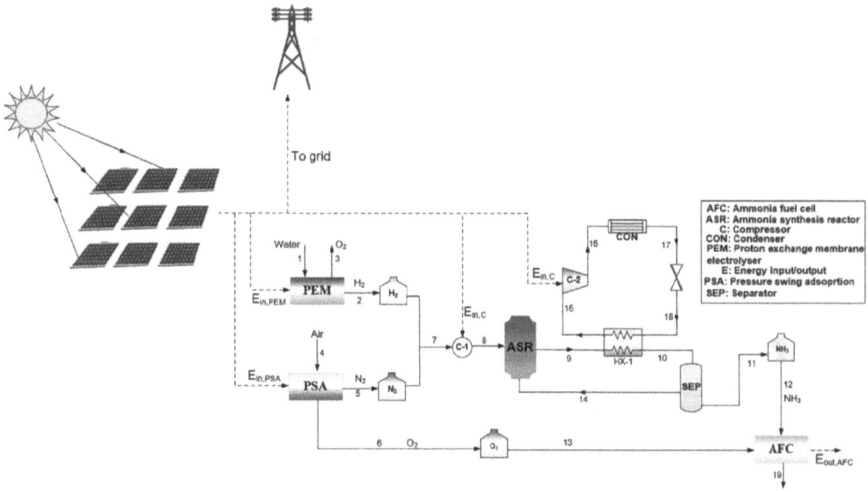

Figure 5.16 The schematic diagram of a solar power plant's integrated ammonia synthesis and fuel cell system.

They suggested that wind turbines be used in combination with solar energy systems to reduce intermittency. Additionally, the developed system's technoeconomic and exergoeconomic evaluations can be conducted to understand its economic performance better.

Galvez et al. [29,30] published an alternative NH_3 synthesis involving two thermochemical reactions in a cyclic method. This noncatalytic method allows for NH_3 synthesis under ambient conditions and without the need for high-intensity H_2 production through steam reforming (bypass). However, since a portion of the system is endothermic and needs a high temperature (greater than 1500°C), additional energy is required that provides heat during the reduction reaction (Fig. 5.17). Galvez et al. suggested using concentrated solar energy to obtain high temperature during the NH_3 synthesis to meet this energy requirement. The proposed method also involves the energy-intensive process of nitrogen removal from the air. However, solar energy utilization has a high degree of uncertainty in achieving the target temperature due to fluctuating conditions such as weather and geographical conditions.

An integrated system for NH_3 synthesis, biomass carbonization, and power generation from an EFB is presented in this report, also adopting the two-step cyclic process. Currently, palm milling waste is growing due to palm oil's rapid expansion, especially in Southeast Asia. Based on the output projection, it is also demonstrated by the annual growth rate of 7.2% from

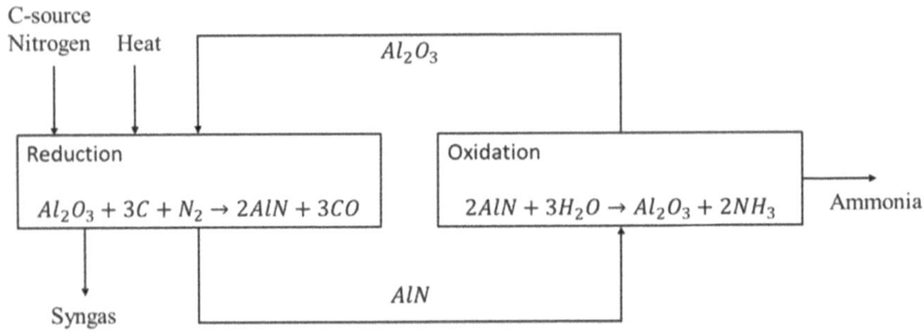

Figure 5.17 Conceptual mechanism of thermochemical cyclic NH_3 production.

2016 to 2021. As a result, high-potential biomass wastes from this sector, including EFB, must be used with suitable technologies to optimize their economic value while minimizing the environmental impacts. Improper disposal methods, such as open burning, can result in many environmental issues. Biomass pretreatment (evaporation and carbonization), combustion, the thermochemical cycle for NH_3 synthesis, and power generation are the key components of the integrated system. The integrated conversion system was developed with the intent of minimizing exergy loss through enhanced process integration. Exergy recovery is performed first to manipulate a single process employing exergy elevation to conduct a self-heat exchange. Furthermore, process integration is carried out to utilize unrecovered heat from one process to assist other processes, thus reducing overall energy destruction in the system.

5.4.1 Proposed system configuration

Fig. 5.18 depicts a simplified diagram of the integrated NH_3 process. It mainly consists of nitrogen generation through PSA, biomass pretreatments (evaporation and carbonization), combustion, NH_3 synthesis (reduction and oxidation), and power generation. Evaporation and pyrolysis (carbonization) are used to obtain coal-like fuel from raw material. The evaporator employs a fluidized bed because of its proper particle mixing, uniform temperature distribution, rapid heat, and moisture transfer. The compressor's evaporation process (in the heat recovery process) and the blower's operation both require electricity. Furthermore, biomass pyrolysis is carried out, which is usually done at temperatures above 300 °C. Pyrolysis is the process of thermal cracking in the absence of oxygen or a low-oxygen environment. It aims to increase the fixed carbon content while also removing moisture and volatiles.

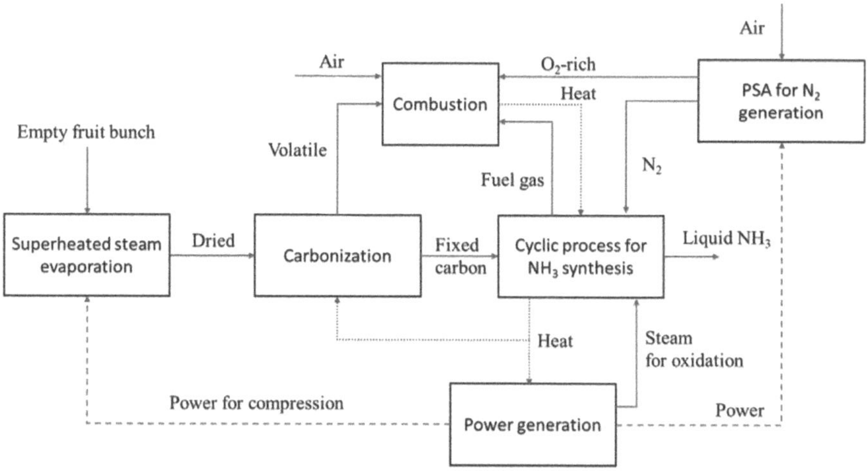

Figure 5.18 Conceptual diagram representing the NH_3 production scheme from empty fruit bunch (EFB).

The pyrolysis gas is used as a fuel in the furnace to provide the majority of the necessary heat throughout the system, especially during the reduction (an endothermic process at a high temperature exceeding 1500°C). Syngas (mostly CO) formed during the oxidation process is also used as fuel in the combustor. The solid carbon is sent to an NH_3 synthesis reactor with high-concentrated nitrogen obtained from the PSA unit. The cyclic process of alumina (Al_2O_3) reduction and aluminum nitride (AlN) oxidation is employed to generate syngas and NH_3 in this process, which facilitates the reduction and oxidation of alumina (Al_2O_3) and aluminum nitride (AlN) in two separate reactors. Oxygen and nitrogen are carried by alumina and aluminum nitride, respectively. Furthermore, the remaining heat is recovered, and high-temperature steam is produced to generate electricity.

5.4.2 Process simulation and analysis methodology

Aspen Plus V8.8 is employed for system simulation, numerical analysis, and optimization of both material and energy balances. However, due to a lack of equilibrium and physical property details, simulation with nonconventional solids, such as biomass feedstock, has limitations. Component Attributes such as PROXANAL (proximate analysis of solid fuel), ULTANAL (ultimate analysis), and SULFANAL are frequently used to characterize nonconventional solids (forms of sulfur analysis). Nonconventional solids are only represented by enthalpy and density models for phase and chemical equilibrium calculations. The enthalpy and density of nonconventional components are estimated using the HCOALGEN and DCOALIGT models, respectively. The IDEAL property approach quantifies the physical properties of conventional mixed elements and CISOLID components in this model.

The composition of EFB is described in Table 5.5. Furthermore, the following assumptions are provided for system boundaries and evaluating system efficiency during process modeling and calculation.

- In steady-state conditions, the proposed system is modeled.
- The temperature is set to 25°C, and the air pressure is set to 101 kPa.
- Both the compressor and the blower have an isentropic efficiency of 87%.
- In the heat exchanger and the whole process reactors, heat losses are negligible.
- The nitrogen and oxygen concentrations in the air are 79 and 21 mol%, respectively.
- The flow rate of raw EFB used in this study is 100 t/h, based on the percentage of EFB derived from a FFB of 20% and collected from 10 palm mills in Indonesia.

In this study, the exergy analysis of a system is performed by modeling exergy balance as follows:

$$\sum_{i} m_{in,i}\ ex_{in,i} = \sum_{e} m_{out,e}\ ex_{out,e} + W + \sum_{i} Ex_{des,\ j} \tag{5.21}$$

Table 5.5 Biomass composition used in this study.

Component	Raw EFB
Proximate analysis	
Fixed carbon (wt% wb)	3.71
Volatile matter (wt% wb)	34.84
Moisture (wt% wb)	60.00
Ash (wt% wb)	1.46
Ultimate analysis	
Carbon (wt% db)	46.62
H_2 (wt% db)	6.45
Oxygen (wt% db)	45.66
Nitrogen (wt% db)	1.21
Sulfur (wt% db)	0.03
Calorific value (MJ/kg)	17.02

where m, ex, W, and Ex are the mass, specific exergy, mechanical or electrical work, and exergy, respectively. The total exergy (ex) is the sum of physical exergy (ex_{ph}) and chemical exergy (ex_{ch}) and calculated as follows:

$$ex = ex_{ph} + ex_{ch} \tag{5.22}$$

$$ex_{ph} = (h - h_0) - T_0(s - s_0) \tag{5.23}$$

$$h - h_0 = \int_{T_0}^{T} C_p dT, \tag{5.24}$$

$$s - s_0 = \int_{T_0}^{T} \frac{C_p}{T}, \tag{5.25}$$

$$ex_{ch} \sum_{e} x_i \left[ex_{ch,i} + RT_0 \ln\left(x_i / \sum x_i \right) \right] \tag{5.26}$$

where T, C_p, h, s, x, and R are temperature, heat capacity, enthalpy, specific entropy, molar fraction, and the universal gas constant, respectively.

The proposed system is built using enhanced process integration in process development. This procedure takes a different approach than the commonly used pinch technology in industrial processes. In order to prevent high exergy destruction, the pinch technology solution improves the system's efficiency by incorporating and optimizing the heat recovery systems. Several articles and books have been written and published to address and explain the pinch analysis for various systems [31–33]. The drawbacks of the pinch technology approach are that it only focuses on the heat exchange and heat recovery network, rather than manipulating the system, such as increasing exergy by compression. Typically, the pinch technology approach is implemented after a thermodynamic investigation of all concerned processes. After that, the content of each stream is listed in order to determine its exergy quality. The composite curves and the heat exchanger network are then analyzed and formulated in the next step. Pinch analysis is a practical application of thermodynamics that is based on simple thermodynamics. It has been implemented to maximize heat recovery and resolve utility schemes at different levels of the industry.

Since pinch technology focuses primarily on temperature control, it is possible to improve the system by modifying it and incorporating other factors such as exergy elevation, electrical components, etc. Process integration and exergy recovery are conducted simultaneously after addressing the system's reactions to obtain high energy efficiency. The enhanced process integration principle is based on exergy recovery via exergy elevation before integrating different processes through a heat exchanger network. As a result, this basic concept differs significantly from standard process integration. The hot stream (the stream in which its exergy is elevated) could be formed as a result of this exergy elevation (process stream, cold stream). As a consequence, an efficient heat pairing is possible. This method is thought to promote a self-heat exchange, thus reducing exergy destruction. The effort to minimize energy consumption in the process has improved, as shown in Fig. 5.19. Overall quality can be enhanced by lowering the exergy destruction rate. As can be identified, most existing systems use conventional integration methods, with just a few pinch technology implementations. Exergy loss/destruction can be greatly reduced by using improved process integration to replace obsolete processes.

5.4.3 Results and analyses of the proposed system based on enhanced process integration

The process flow representation for the proposed NH_3 processing from EFB is shown in detail in Fig. 5.20. Before being heated, raw EFB is first shredded (HX1). After that, it is dried in the main evaporator (HX2) with a target moisture content of 5 wt% wb. The required drying process is 142°C, according to the numerical estimation for equilibrium moisture

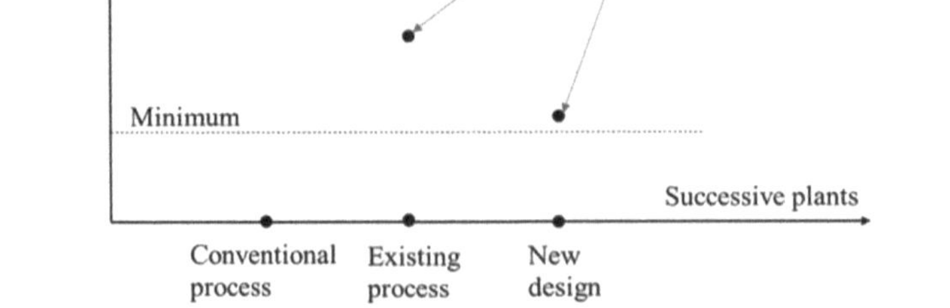

Figure 5.19 Improvement of minimizing exergy destruction.

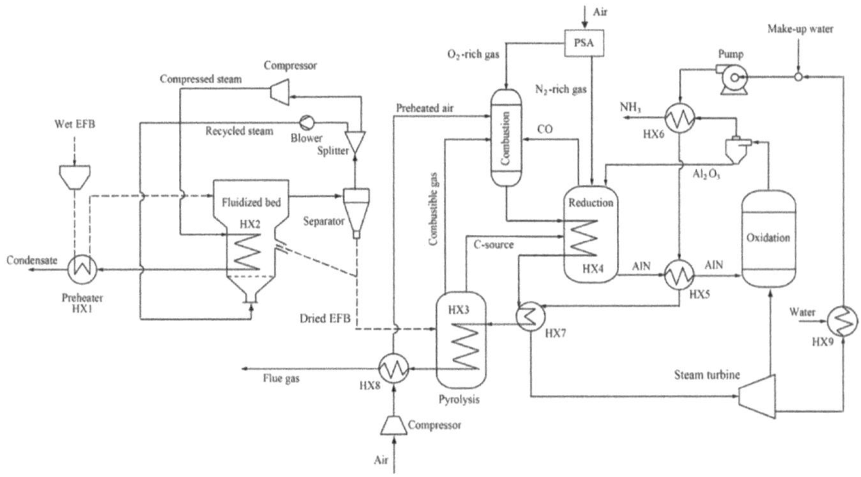

Figure 5.20 Process flow diagram of NH_3 production from the empty fruit bunch (EFB).

content. Because of its excellent efficiency, such as rapid heat transfer and uniformly distributed drying quality, a fluidized bed is utilized in the main evaporator. The vapor that exits the evaporator is separated into two streams: fluidizing steam (which is reused during drying) and heat-source steam (finally purged after being compressed and recovered for its heat). The heat-source steam is then passed through a compressor to increase its exergy rate before being employed as a heat source in HX2. Between the heat of EFB's water evaporation and the heat of compressed steam condensation, the heat transfer of latent heat is substantially dominant in the main evaporator. A self-heat exchange among the streams in the same evaporation module can help facilitate the evaporation process, reducing exergy loss.

The energy consumed during evaporation is used for compressor and blower operations. Table 5.6 describes the evaporation performance. The solid EFB is transferred to the pyrolysis reactor for carbonization using a downdraft fixed bed type with indirect heating after being dried in the fluidized bed reactor. The residual high temperature of flue gas (750°C) from the burning process is used to conduct this stage at a temperature of 500°C (HX3). Because of the feedstock's high volatile content, decomposition of EFB takes place effectively at this stage; thus, combustible gases (such as H_2 and carbon monoxide) and solid yields are produced and used as fuel for subsequent processes.

Table 5.6 Detailed performance of evaporation process.

Properties	Value
General conditions	
Mass flow of EFB (t/h)	100
Initial moisture content (wt% wb)	60.00
Target moisture content (wt% wb)	5.00
Minimum temperature approach of heat exchanger (°C)	10
Compressor	
The pressure at the compressor outlet (kPa)	212
The temperature at the compressor outlet (°C)	236
Compressor work (MW)	2.92
Steam condensation temperature (°C)	122
Main evaporator (HX1)	
Bed temperature (°C)	142
Evaporation duty (MW)	39.87

The pyrolyzed gases and solid carbon are also delivered to the combustion reactor and reduction reactor (in the thermochemical cycle), respectively. The solid carbon is sent to an NH_3 synthesis reactor with high-concentrated nitrogen obtained from the PSA unit. The cyclic process of alumina (Al_2O_3) reduction and aluminum nitride (AlN) oxidation is used to generate syngas and NH_3 in this process, which involves the reduction and oxidation of alumina (Al_2O_3) and aluminum nitride (AlN) in two separate reactors. Table 5.7 depicts the main parameters of the thermochemical cycle, which consists of reduction and oxidation reactions. Furthermore, the thermochemical cycle's residual heat is recovered and used to generate steam because of the high-temperature operation, which is then used to generate electricity (HX5, HX6, and HX7).

In order to enhance total performance, it is important to minimize exergy loss when designing the process. Exergy loss or destruction can be divided into primary and secondary exergy destruction presented in this report. The primary destruction is the loss of exergy that occurs as the phase or reaction progresses. The secondary destruction is lost during processes that support the main reaction, such as losses during heat exchange, work, or heat loss. The irreversibility of the overall process due to heat exchange

Table 5.7 The detailed performance of the thermochemical cycle.

Properties	Value
General	
Mass flow of solid carbon (t/h)	12.40
Nitrogen supply (t/h)	15.00
Circulated Al_2O_3 (t/h)	36.28
Reduction	
Reduction pressure (kPa)	101.32
Reduction temperature (°C)	1500
Reaction type	Endothermic
Reactor type	Counter-current moving bed reactor
Oxidation	
Oxidation temperature (°C)	950, 1000, 1100, 1200
Oxidation pressure (kPa)	101.32
Reactor type	Entrained flow reactor
Reaction type	Exothermic

increases exergy degradation, lowering energy efficiency. Fig. 5.21 depicts the predominant exergy loss due to heat exchanges. Most exergy is lost due to unrecovered heat in the flue gas and NH_3 + steam leaving the integrated system.

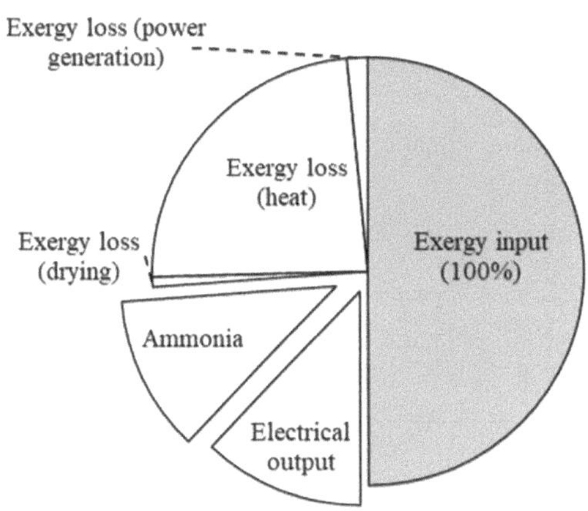

Figure 5.21 Overall exergy balance.

5.4.4 Performance of thermochemical cycle

At reduction and oxidation temperatures of 1500 and 1000°C, respectively, Fig. 5.22 depicts the effect of carbon reduction efficiency on the amount of NH_3 released during the thermochemical cycle. According to experimental studies, oxidation at 1000°C generates the maximum NH_3 yield, which decreases as the temperature rises. Many factors affect the fractional coefficient of carbon reduction, including pressure, equilibrium composition, time, and the different kinds of carbon sources. Many carbon sources have been studied for their kinetics performance during a reduction in the thermochemical cycle, including wood charcoal, active carbon, black rubber, and petroleum coke. To fully comprehend the reduction behavior of pyrolyzed EFB at 500 °C, more investigation is required. The fractional reduction values (0.75, 0.80, 0.85, and 0.90, and 0.95) are simulated since the reduction of pyrolyzed EFB is closed to the studied carbon sources due to material similarity. As shown in Fig. 5.23, the efficiency of NH_3 production ranges from 19.15% to 24.70% during reduction and oxidation reactions at 1500 and 1000°C, respectively. These findings are reasonable given that the EFB fuel has a high volatile matter content that is pyrolyzed during carbonization at 500°C.

The effect of oxidation temperature on NH_3 output during the thermochemical cycle is depicted in Fig. 5.24. The fractional reduction of carbon is set to 0.9 to examine the effect of oxidation temperature on system performance, especially the NH_3 yield. According to the experimental results, oxidation at 1000°C produces the maximum NH_3 yield. By increasing the oxidation temperature from 950 to 1000°C, the NH_3 yield

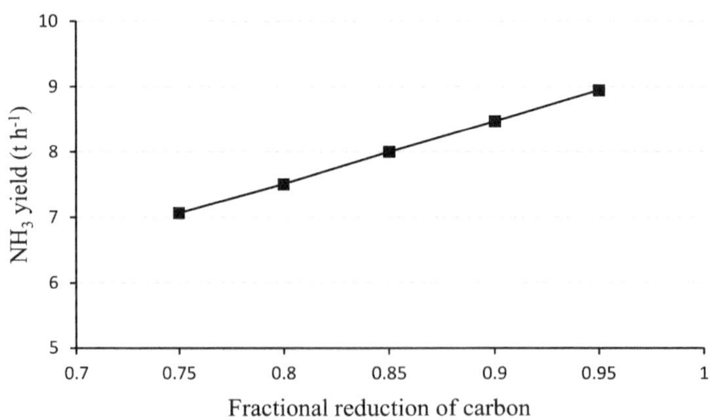

Figure 5.22 The effect of carbon reduction performance on NH_3 produced.

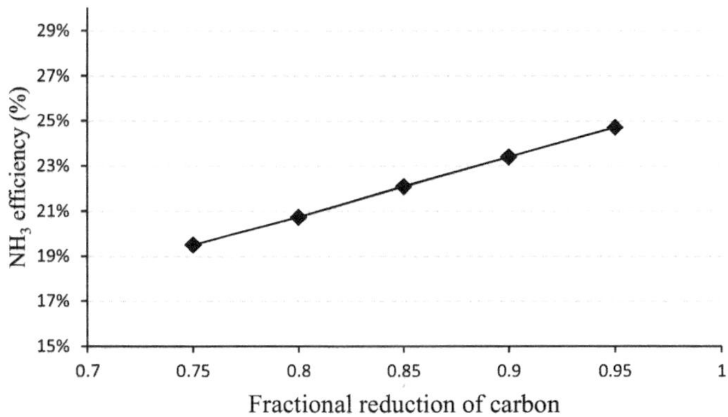

Figure 5.23 The effect of carbon reduction performance on NH_3 efficiency.

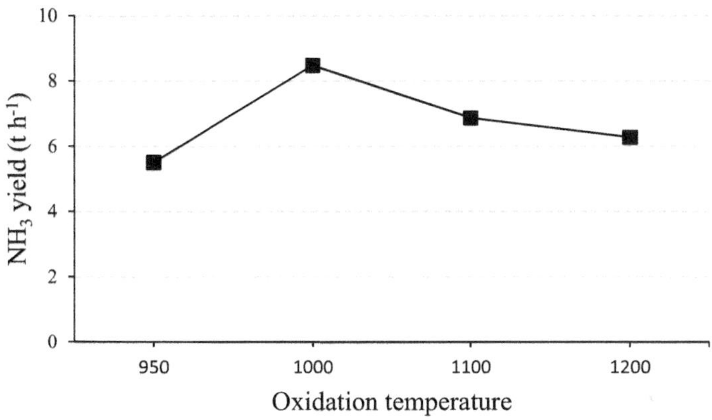

Figure 5.24 The effect of oxidation temperature on NH_3 produced.

increases from 5.5 to 8.5 t/h, and the NH_3 performance rises from 15.18% to 23.39%. As the oxidation temperature exceeds 1000°C, however, the NH_3 yield decreases.

5.4.5 Performance of power generation system

Reduction and oxidation temperatures are set at 1500 and 1000°C, respectively, to understand power generation efficiency better. Due to the high temperature during the thermochemical cycle, three heat exchangers are used in a Rankine cycle system to recover available heat and produce steam for power generation. Fig. 5.25A depicts the effect of steam turbine inlet pressure on net power and output vapor efficiency. The temperature

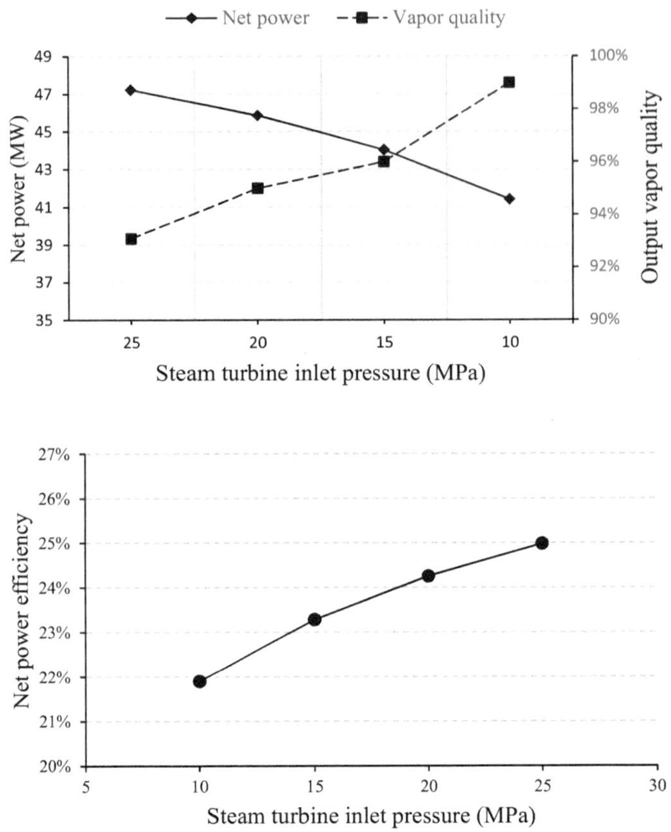

Figure 5.25 Effect of inlet pressure of steam turbine on (A) net power and vapor quality and (B) net power efficiency (η_{power}).

inlet of the steam turbine is set to 600°C. The results indicate that raising the inlet pressure from 10 to 25 MPa increases net power from 41.41 to 47.24 MW. As seen in Fig. 5.25B, this corresponds to an improvement in power efficiency from 21.89% to 24.98%. The steam turbine's output vapor efficiency drops from 99% to about 93% as the inlet pressure rises. Water usage for power generation is about 50 kg/s. However, a rise in inlet pressure has no impact on the efficiency of NH_3 output (η_{NH_3}).

The effect of the steam turbine's inlet temperature on power efficiency (η_{power}) is visualized in Fig. 5.26. We set the inlet pressure to 20 MPa to investigate the effect of the steam turbine's inlet temperature. After that, the quantity of water is adjusted to achieve the desired temperature. According to the results, the inlet temperature of 650°C has the highest overall

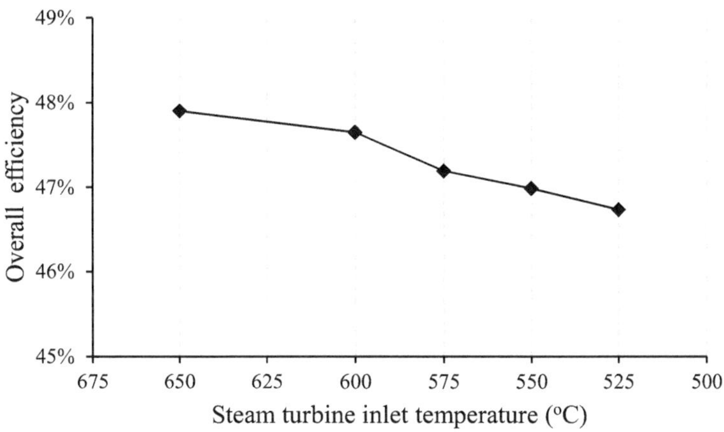

Figure 5.26 Effect of inlet temperature of the steam turbine on the total efficiency.

efficiency (η_{total}) of about 47.89%, with a net power produced of 46.35 MW. When the inlet temperature is reduced to 525°C, the overall efficiency drops to 46.73%.

In conclusion, enhanced process integration is employed to model the novel integrated system for NH_3 production. Evaporation, carbonization, the thermochemical cycle, and power generation are the integrated system's key elements. Some parameters during the thermochemical cycle and power generation were investigated to understand the system performance better. In the thermochemical cycle, nitrogen, carbon source, steam, circulated alumina (Al_2O_3) particles, and aluminum nitride(AlN) undergo reduction and oxidation in a cyclic method. Heat recovery could be used to produce power from the Rankine cycle since the cyclic process occurs at high temperatures. The results demonstrate that the integrated system could effectively coproduce power and NH_3 with a total net efficiency of around 48%. The use of 100 t/h of EFB could generate 8.95 t/h of NH_3 and 46.35 MW of electricity, respectively.

References

[1] Zion Market Research. Palm oil market analysis by derivative (CPO, PKO, PKC, and others), and for edible oil, cosmetics, bio-diesel, lubricants, surfactants and other applications - global industry perspective, comprehensive analysis, and forecast 2015–2021. Maharashtra. 2016.

[2] UNDP. Indonesia palm oil. 2018. p. 1–33.

[3] Haafiz MKM, Hassan A, Zakaria Z, Inuwa IM. Isolation and characterization of cellulose nanowhiskers from oil palm biomass microcrystalline cellulose. Carbohydr Polym 2014;103:119–25. https://doi.org/10.1016/j.carbpol.2013.11.055.
[4] Mukherjee I, Sovacool BK. Palm oil-based biofuels and sustainability in southeast Asia: a review of Indonesia, Malaysia, and Thailand. Renew Sustain Energy Rev 2014;37:1–12. https://doi.org/10.1016/j.rser.2014.05.001.
[5] Embrandiri A, Ibrahim MH, Singh RP. Palm oil mill waste utilization: sustainability in Malaysian context palm oil mill wastes utilization; sustainability in the Malaysian context. Int J Sci Res Publ 2016;3:1–7.
[6] Heredia Salgado MA, Tarelho LAC, Matos MAA, Rivadeneira D, Narváez CRA. Palm oil kernel shell as solid fuel for the commercial and industrial sector in Ecuador: tax incentive impact and performance of a prototype burner. J Clean Prod 2019;213:104–13. https://doi.org/10.1016/j.jclepro.2018.12.133.
[7] Yusoff S. Renewable energy from palm oil - innovation on effective utilization of waste. J Clean Prod 2006;14:87–93. https://doi.org/10.1016/j.jclepro.2004.07.005.
[8] Geng A. Conversion of oil palm empty fruit bunch to biofuels. Liq. Gaseous solid biofuels - convers. Tech. InTech; 2013. https://doi.org/10.5772/53043.
[9] McDowall W, Eames M. Forecasts, scenarios, visions, backcasts and roadmaps to the hydrogen economy: a review of the hydrogen futures literature. Energy Policy 2006;34:1236–50. https://doi.org/10.1016/j.enpol.2005.12.006.
[10] Zhao Y, Lee DH, Seok MY, Lee JA, Phaniraj MP, Suh JY, et al. Resistance of CoCrFeMnNi high-entropy alloy to gaseous hydrogen embrittlement. Scripta Mater 2017;135:54–8. https://doi.org/10.1016/j.scriptamat.2017.03.029.
[11] Goshome K, Yamada T, Miyaoka H, Ichikawa T, Kojima Y. High compressed hydrogen production via direct electrolysis of liquid ammonia. Int J Hydrogen Energy 2016;41:14529–34. https://doi.org/10.1016/J.IJHYDENE.2016.06.137.
[12] Klerke A, Christensen CH, Nørskov JK, Vegge T. Ammonia for hydrogen storage: challenges and opportunities. J Mater Chem 2008;18:2304–10. https://doi.org/10.1039/b720020j.
[13] Mizuno Y, Ishimoto Y, Sakai S, Sakata K. Economic analysis on international hydrogen energy career supply chains. J Jpn Soc Energy Resour 2017;38:11–7.
[14] Wijayanta AT, Oda T, Purnomo CW, Kashiwagi T, Aziz M. Liquid hydrogen, methylcyclohexane, and ammonia as potential hydrogen storage: comparison review. Int J Hydrogen Energy 2019;44:15026–44. https://doi.org/10.1016/j.ijhydene.2019.04.112.
[15] Agency for Natural Resources and Energy. Basic hydrogen strategy. 2017.
[16] Aasadnia M, Mehrpooya M. Large-scale liquid hydrogen production methods and approaches: a review. Appl Energy 2018;212:57–83. https://doi.org/10.1016/j.apenergy.2017.12.033.
[17] Japan Long Course Ferry Service Association. Fuel oil price change chart. 2018.
[18] Aziz M, Oda T, Morihara A, Kashiwagi T. Production of ammonia from low rank coal employing chemical looping and haber-bosch process. Chem Eng Trans 2017;61:79–84. https://doi.org/10.3303/CET1761011.
[19] Reese M, Marquart C, Malmali M, Wagner K, Buchanan E, McCormick A, et al. Performance of a small-scale haber process. Ind Eng Chem Res 2016;55:3742–50. https://doi.org/10.1021/acs.iecr.5b04909.
[20] Andersson J, Lundgren J. Techno-economic analysis of ammonia production via integrated biomass gasification. Appl Energy 2014;130:484–90. https://doi.org/10.1016/j.apenergy.2014.02.029.
[21] Arora P, Hoadley AFA, Mahajani SM, Ganesh A. Multi-objective optimization of biomass based ammonia production - potential and perspective in different countries. J Clean Prod 2017;148:363–74. https://doi.org/10.1016/j.jclepro.2017.01.148.

[22] Gilbert P, Alexander S, Thornley P, Brammer J. Assessing economically viable carbon reductions for the production of ammonia from biomass gasification. J Clean Prod 2014;64:581−9. https://doi.org/10.1016/j.jclepro.2013.09.011.
[23] Sivasangar S, Zainal Z, Salmiaton A, Taufiq-Yap YH. Supercritical water gasification of empty fruit bunches from oil palm for hydrogen production. Fuel 2015;143:563−9. https://doi.org/10.1016/J.FUEL.2014.11.073.
[24] Sivasangar S, Taufiq-Yap YH, Zainal Z, Kitagawa K. Thermal behavior of lignocellulosic materials under aerobic/anaerobic environments. Int J Hydrogen Energy 2013;38:16011−9. https://doi.org/10.1016/j.ijhydene.2013.09.083.
[25] Ahmad AL, Sumathi S, Hameed BH. Coagulation of residue oil and suspended solid in palm oil mill effluent by chitosan, alum and PAC. Chem Eng J 2006;118:99−105. https://doi.org/10.1016/j.cej.2006.02.001.
[26] Fan LS. Chemical looping systems for fossil energy conversions. NJ, USA: John Wiley & Sons, Inc.; 2010. https://doi.org/10.1002/9780470872888.
[27] Sajid MU, Bicer Y. Thermodynamic assessment of chemical looping combustion and solar thermal methane cracking-based integrated system for green ammonia production. Therm Sci Eng Prog 2020;19:100588. https://doi.org/10.1016/j.tsep.2020.100588.
[28] Siddiqui O, Dincer I. A new solar energy system for ammonia production and utilization in fuel cells. Energy Convers Manag 2020;208:112590. https://doi.org/10.1016/j.enconman.2020.112590.
[29] Gálvez ME, Hischier I, Frei A, Steinfeld A. Ammonia production via a two-step Al_2O_3/AlN thermochemical cycle. 3. Influence of the carbon reducing agent and cyclability. Ind Eng Chem Res 2008;47:2231−7. https://doi.org/10.1021/ie071244w.
[30] Gálvez ME, Frei A, Halmann M, Steinfeld A. Ammonia production via a two-step Al_2O_3/AlN thermochemical cycle. 2. Kinetic analysis. Ind Eng Chem Res 2007;46:2047−53. https://doi.org/10.1021/ie061551m.
[31] Kemp I. Pinch analysis and process integration. Elsevier Ltd; 2006. https://doi.org/10.1016/B978-0-7506-8260-2.X5001-9.
[32] Higa M, Freitas AJ, Bannwart AC, Zemp RJ. Thermal integration of multiple effect evaporator in sugar plant. Appl Therm Eng 2009;29:515−22. https://doi.org/10.1016/j.applthermaleng.2008.03.009.
[33] Aspelund A, Berstad DO, Gundersen T. An Extended Pinch Analysis and Design procedure utilizing pressure based exergy for subambient cooling. Appl Therm Eng 2007;27:2633−49. https://doi.org/10.1016/j.applthermaleng.2007.04.017.

CHAPTER 6

Integrated systems from agricultural waste for power generation

Arif Darmawan[1], Muhammad Aziz[2], Muhammad Kunta Biddinika[3] and Koji Tokimatsu[4]

[1]Agency for the Assessment and Application of Technology (BPPT), Puspiptek Serpong, Tangerang Selatan, Indonesia; [2]Institute of Industrial Science, The University of Tokyo, Meguro-ku, Tokyo, Japan; [3]Faculty of Industrial Technology, Universitas Ahmad Dahlan, Indonesia; [4]Department of Transdisciplinary Science and Engineering, Tokyo Institute of Technology, Midori-ku, Yokohama, Kanagawa, Japan

This chapter aims to address the use of readily available biomass waste from agricultural activities as a cost-effective alternative fuel. In this scenario, agricultural waste is no longer openly burned or left to rot on the fields. It has contaminated the atmosphere and released greenhouse gases (GHGs) for a long time. The concern here is figuring out how to use waste with a low thermal value. Agricultural biomass waste refers to organic materials that do not go directly into foods or other items and are generally found in the form of crop residue stalks, leaves, roots, seeds, seed shells, and other similar materials. Wheat and paddy straw, bagasse (residue from sugarcane milling), seed hulls (rice husk, groundnut husk), waste wood from timber processing, waste from animal husbandry, and other examples are common agricultural waste [1]. Several technologies to utilize biomass, including agricultural waste, have been established over the years. In order to generate liquid and/or gaseous fuels, both biochemical (fermentation) and thermochemical (pyrolysis, gasification) approaches can be used. These fuels can be used directly as sources of energy or further converted into useful chemicals and products. Solid fuels can also be created by physical processes such as briquetting and compaction. However, since microbes have difficulty degrading the highly crystalline cellulose embedded in the lignin matrix, the biochemical conversion is considered not appropriate for lignocellulosic biomass breakdown [2]. Although a high-intensity heat source is needed, the thermochemical process is still considered an effective technology for converting lignocellulosic biomass into bioenergy. Thermochemical conversion is described as the chemical dissociation of biomass at high temperatures, which includes bond breaking and the transformation of biomass into bio-char (solid), bio-gas, or syngas bio-oil (liquid) [3].

Innovative Energy Conversion from Biomass Waste
ISBN 978-0-323-85477-1
https://doi.org/10.1016/B978-0-323-85477-1.00003-8

6.1 Integrated system of rice production and electricity generation

Diversification of energy sources, especially renewables, is critical for satisfying energy demand sustainably and properly while still addressing environmental concerns. Renewables include wind, biomass, solar, industrial and agricultural wastes, geothermal energy, and hydropower. Rice plantations, one of the biomass resources, have recently grown rapidly as a primary food source in many countries, especially in Asia. Rice plantation waste, consisting of rice straw and husk, amounts to around 1,370,000 million t per year. These products account for the bulk of total rice plantation products and have a relatively high economic value. 0.28 kg of rice husk and 1.1 kg of rice straw are produced for every kg of rice grain produced [4,5]. This situation causes a slew of issues related to improper disposal activities of rice waste products, such as open burning during peak harvesting season, which pollutes the environment. As a result, advanced rice husk and straw use, including energy harvesting, is critical from both an economic and environmental standpoint. Around 90% of the world's rice commodity is produced in Asian countries, with Southeast Asia accounting for 29%. From the 19 million t of rice husks generated each year, it was estimated that Southeast Asia's > 100,000 rice mills could generate 16,720 MW of electricity [6].

At the industrial level, effective energy conservation is crucial. Several technologies, such as waste heat recovery, steam efficiency improvement, insulation quality improvement, and the advancement of alternative technology, can increase energy efficiency of rice production processes. Each technology has its own set of advantages, opportunities, and challenges, as well as various stages of development and deployment. Researchers have explored and proposed various drying methods for rice grain, including hot air drying, vacuum drying, superheated steam drying (SSD), and fluidized bed drying, as well as the parboiling process for the water removal process. The parboiling process, which involves soaking paddy until it is saturated (about 30% moisture on a wet basis), steaming to gelatinize the starch, and then drying, can have a major impact on paddy quality (Fig. 6.1). Unfortunately, the parboiling process has a number of energy-related drawbacks, including higher upfront costs, additional drying costs, time constraints, and environmental concerns [7]. As a result, a number of studies aimed at replacing this process have been conducted. Due to higher drying speeds, better efficiency than traditional drying, lower net energy consumption due

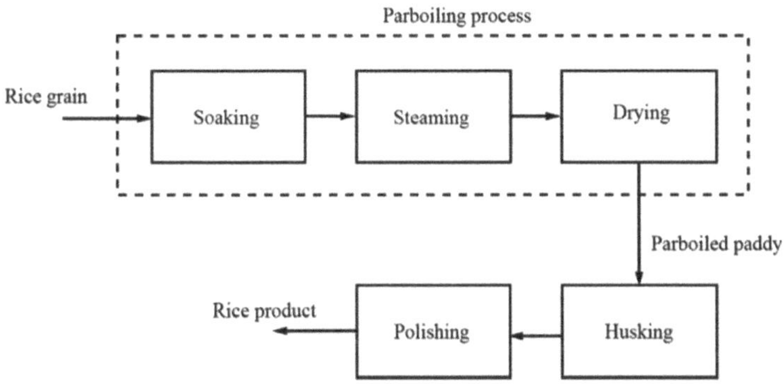

Figure 6.1 Steps of rice processing in modern mills, including the parboiling process.

to the possibility of better heat recovery, and the avoidance of fire and explosion hazards make the SSD process a promising solution. Rice grain drying using superheated steam as an alternative to the parboiling process was investigated by Chaiyong et al. [8] and Wu et al. [9]. However, they were only interested in the impact of SSD on yield characteristics of dried rice grain. No effort has yet been made to increase the system's overall energy efficiency.

Compared to rice straw, electricity production from rice husk has advanced significantly in terms of energy recovery. Suramaythangkoor et al. investigated the feasibility of rice straw for power generation and found it to be beneficial in terms of both cost and reduction of GHG emissions [10]. Several recent studies have proposed and evaluated rice waste utilization systems for energy production, especially power generation. Rice husk and straw thermochemical and biochemical conversions have previously been widely studied. Rice straw and/or rice husk gasification was identified as a promising energy conversion technology due to its high conversion efficiency among the technologies examined. Yoon et al. used rice husk and pellets in a fixed-bed gasification reactor to generate electricity [11]. They claimed to be able to achieve a cold gas efficiency of about 70%. Calvo et al. [12] have used a fluidized bed to conduct rice straw gasification. They discovered that using fluidized bed gasification to produce high-quality syngas with a conversion efficiency of up to 61% could produce high-quality syngas. Unfortunately, their studies focused exclusively on the gasification system rather than designing an integrated power generation system.

Furthermore, Prabowo et al. [13] and Chiang et al. [14] looked into rice straw gasification in both downdraft and updraft gasifiers. The former investigated the use of CO_2 as an alternative to steam as a gasifying agent in rice straw gasification to improve thermal efficiency. They discovered that using CO_2 as a gasifying agent has a high potential for increasing the conversion process's thermal efficiency (by up to 60% at 950°C gasification temperature). Regrettably, the studies above primarily concentrated on the conversion phase (gasification) rather than designing a fully integrated process capable of achieving high total energy efficiency. Almost no studies are available in the literature considering the requirements for effectively combining rice production and power generation, especially in terms of total energy efficiency. In this chapter, a novel integrated system consisting of SSD for the rice grain, husking, torrefaction, steam gasification of rice straw and husk, and power generation is discussed based on exergy recovery and process integration. The objective is to minimize the exergy destruction throughout the integrated system to realize high total energy efficiency.

Fig. 6.2 depicts a schematic diagram of the suggested integrated system for rice production and electricity generation. The wet raw rice grain is first fed into the SSD system to remove moisture before being purified in the husking and polishing system. Before being introduced to the gasification system for syngas production, the by-products, rice husk and straw, are torrefied at 200–300°C to boost their physical and chemical feedstock properties. After that, the syngas is sent to be used in the generation of electricity.

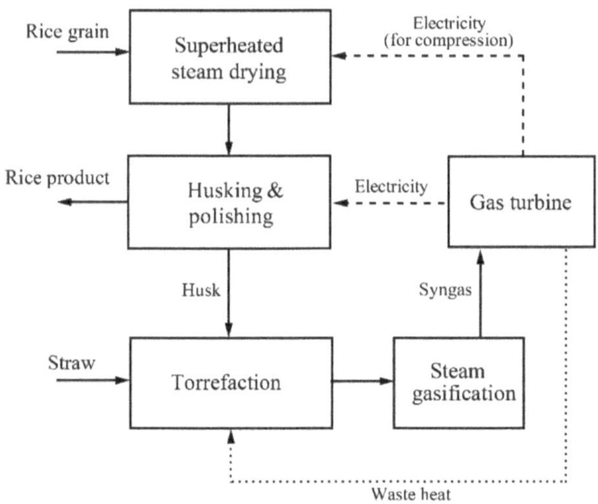

Figure 6.2 Conceptual diagram of the proposed integrated system of rice production and electricity generation.

Rice grain drying, husking, and polishing are the major elements in the rice production process. After harvesting, the raw rice grain is fed into the SSD system (which replaces the parboiling process) to extract moisture. Rice with similar viscoelasticity and textural properties to parboiled rice can be generated using this method. To avoid grain discoloration, mold growth, and germination, the harvested rice grain should immediately be dried. Due to many factors such as physical composition, structure, and the process environment, drying or evaporation is considered a complex process. Inside the particles, drying involves both free and bound water. In the milling or husking modules, the dried grain is purified by extracting the husk and rice bran (outer layer). The following mass fractions are generated due to these processes: 78% milled rice and 20%–221% husk. After scraping the husk, the bran is removed, resulting in a small reduction in the milled rice fraction from 78% to 68%–72%. Torrefaction (pretreatment), steam gasification, and the gas turbine are the electricity generation system's core processes. Rice husk and rice straw by-products are torrefied between 200 and 290°C, using the heat of flue gas from power generation. The torrefied straw and husk are then placed into the gasification unit. They are transformed at high-temperature condition into syngas, which is mainly composed of fuel gases such as hydrogen and carbon monoxide. Drying, pyrolysis, volatilization, combustion, and char gasification reactions take place within the gasifier, generating syngas. The sulfur and particulate matter are then extracted from the syngas. The syngas is cleaned before being transferred to the combustor, which produces high-temperature, high-pressure gas for gas turbines. The residual heat can also be recovered for the torrefaction operation since the temperature of the flue gas exhausted from the gas turbine is relatively high.

6.1.1 Process modeling and analysis

Table 6.1 summarizes rice husk and straw composition, including ultimate and proximate analyses [15]. Rice husk and straw biomass also have a low sulfur content, so they have a lower propensity to produce SO_2 and H_2S gas [16,17]. According to a study of various rice mills, the average total amount of rice grain processed in one mill is about 200 t/d. Rice grain has a specific heat of 1109 ± 45 MJ/kg/K, and moisture content ranges from 10 to 17 wt% wb. A simulator in Aspen Plus V8.8 was used to model and calculate the proposed integrated system's material and energy balances. The raw materials were classified as nonconventional solids that could not be reacted with other reactants directly. After that, a FORTRAN

Table 6.1 Composition of materials used in the study.

Component	Husk	Straw
Ultimate analysis		
Hydrogen (wt% db)	4.75	5.2
Oxygen (wt% db)	35.47	36.26
Nitrogen (wt% db)	0.52	0.87
Sulfur (wt% db)	0.05	0.18
Carbon (wt% db)	38.83	38.24
Chlorine (wt% db)	0.12	0.58
Ash (wt% db)	20.26	18.67
Proximate analysis		
Volatile (wt% db)	63.52	65.47
Fixed carbon (wt% db)	16.22	15.86
Ash (wt% db)	20.26	18.67
Calorific value (MJ/kg)	15.84	15.09

subroutine was used to decompose solid materials into H_2O, ash, and O_2, H_2, and carbon elements using an RYield block. MIXCINC was the stream class in Aspen Plus, and IDEAL was set as the property method.

6.1.1.1 Superheated steam drying

A schematic diagram of the proposed rice grain drying and a process flow diagram based on exergy recovery are shown in Fig. 6.3. Hot soaking, drying, and steam superheating are the three phases of the presented drying process. The compressed vapor flowing out of the fluidized bed dryer heats the rice grain entering the hot soaking stage (HX1) to the desired temperature. In this state, heat exchange takes place between wet rice grain as the cold stream and condensed steam as the hot stream. The wet rice grain then goes on to the drying stage (HX2), which is carried out in a fluidized bed dryer. The heat of condensation from the compressed steam is exchanged with the moisture of the rice grain. Some parameters, such as temperature and moisture content, are set based on experiments to ensure the rice yield's consistency. SSD was carried out in this study at a bed temperature of 150°C and a target moisture content of 18 wt% wb.

The steam from the main dryer is divided into recycled steam (F10) and compressed steam (F7). The water extracted from the soaked rice grain is equal to the quantity of compressed steam. Since steam (F6) is compressed to increase exergy level, steam's physical properties, including its saturation

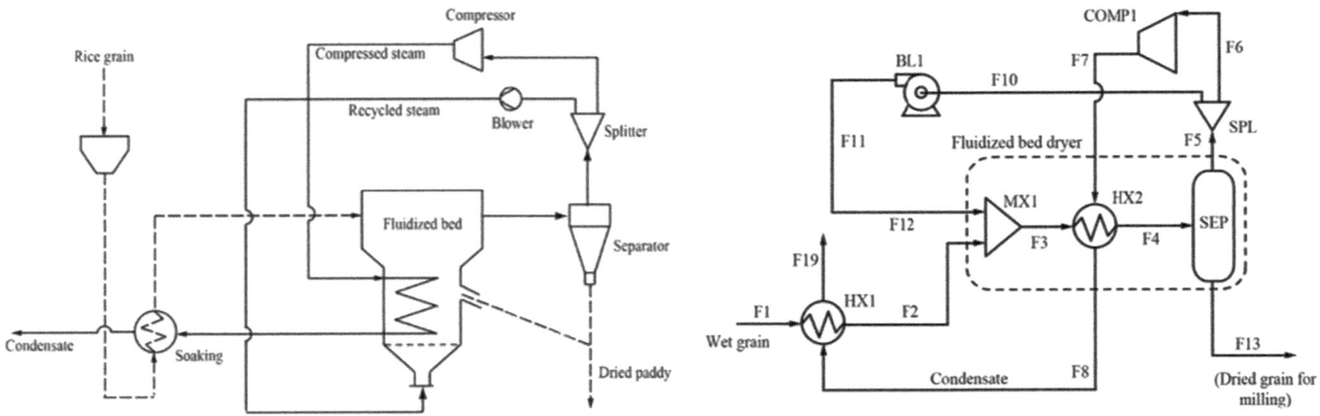

Figure 6.3 Schematic diagram of the proposed rice grain drying.

Table 6.2 Drying conditions for superheated steam drying (SSD) of rice grain in the study.

Properties	Value
Average particle diameter (mm)	3.5
Moisture content before soaking (wt% wb)	26
Moisture content after soaking (wt% wb)	60
Moisture content after SSD (wt% wb) [18]	18
Heat capacity (J/kg/K)	1109 ± 45M
Bed temperature (°C) [8]	150
Minimum fluidization velocity (m/s)	2.6
Fluidized bed	
Side length (m)	2
Bed height (m)	2

point, raising the chance of latent heat recovery. Using these steam flows can enhance the drying process. During the operation, the flow rate of recycled steam is set based on an experimental study, ensuring excellent direct contact with rice grain. Table 6.2 lists the assumed conditions throughout the SSD stage. Physical composition, structure, and drying environment are all factors that influence the drying state. Drying often necessitates more energy to resolve many forces, such as chemical bonding and sorption forces. It leads to equilibrium moisture content, which is the balanced condition of active activity of hygroscopic materials. Based on Taechapairoj et al. rice grain's equilibrium moisture content under the superheated steam condition can be estimated [8].

6.1.1.2 Husking and polishing processes

In the postproduction of rice, husking and polishing are important. The methods are used to eliminate the husk and brand layers. Husk and brand extraction can be done in one stage in the modern milling process. Dehusking, aspirating the husk, separating paddy and brown rice, polishing to extract the bran, and grading rice and broken grains are all steps in the process [19]. The following equation is used to calculate the net energy consumption (NEC_{mill}) of the milling process in this report. Both specific energy consumption and electrical energy utilization index are based on values reported by Goyal et al. (5.76 kWh/t and 0.61, respectively) [19,20].

$$NEC_{mill}(KWh) = \frac{\text{Capacity of paddy } (t\, h^{-1})\ x \text{ time of operation } (h)}{\text{Specific Energy Consumption } (KWh\, t^{-1})} \tag{6.1}$$

Therefore, the total electricity consumption, P_{mill}, can be obtained as follows:

$$P_{mill} = \frac{NEC_{mill}}{\text{Electrical Energy Utilization Index}} \tag{6.2}$$

6.1.1.3 Proposed integrated system for torrefaction, steam gasification, and power generation

The conceptual model of the overall integrated power generation system introduced in this study is depicted in Fig. 6.4. Torrefaction, gasification, gas cleaning, and power generation are the primary parts of the integrated process. In order to increase biomass quality and extract moisture, the feedstock is fed to a torrefaction reactor. The torrefied feedstock is then supplied to the gasification module for efficient carbon conversion to fuel gases such as hydrogen and carbon monoxide. The syngas is formed after pyrolysis, volatilization, combustion, and char gasification reactions take place in the gasifier.

Fig. 6.5 depicts a torrefaction and gasification process simulation. The degree of torrefaction is a function of temperature and is calculated using evidence from experimental work in the literature by Chen et al. [21]. They examined the torrefaction of rice waste and focused on the impact of temperature on the torrefied products. Torrefaction, material decomposition, and the gasification reactor were all simulated in the Aspen Plus simulator as RStoic, RYield, and RGibbs blocks, respectively. The RGibbs block was also used in the combustor stage to react carbon with oxygen (in the fed air). Since husk and straw are nonconventional solids, they cannot be reacted with other components directly. The solid materials were then decomposed into water, ash, and elements such as oxygen, hydrogen, and carbon using an RYield block with a FORTRAN subroutine. Some of the solid carbon was reacted with air to provide heat for gasification. The decomposed components were then added to an RGibbs block to conduct the chemical reactions. The heat of conversion of the RYield of BL decomposition, the RGibbs of gasification reaction, and the RGibbs of combustion for BL conversion were equal to the amount of the heat of conversion of the RYield of BL decomposition, the RGibbs of gasification reaction, and the RGibbs of combustion.

In addition, hot syngas with a high thermal energy content was used to produce steam. The temperature of the syngas before cleaning was held above 300°C to prevent gas condensation. After that, the syngas had to be

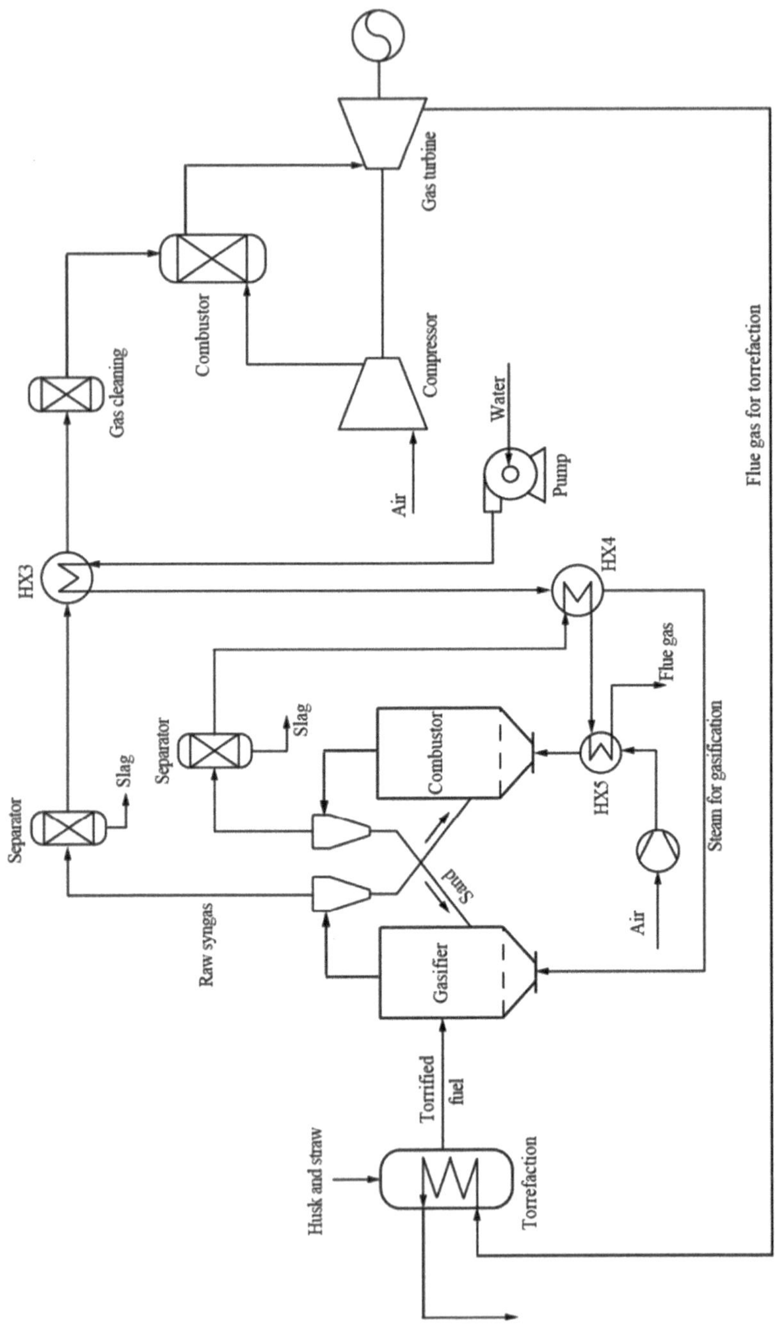

Figure 6.4 Process flow diagram of the proposed system including rice waste torrefaction, gasification, and the gas turbine.

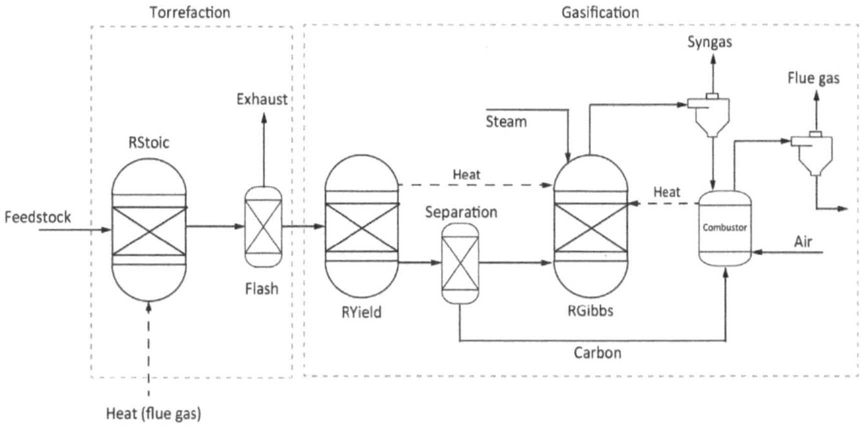

Figure 6.5 Process simulation of torrefaction and steam gasification.

cleaned before it could be pumped into the combustor. During combustion, the hot, high-pressure gas expanded in the gas turbine, generating power. The calculation conditions for the gasification and combined cycle processes are shown in Table 6.3. To improve fluidization and gas transfer within the gasifier, sand (particle size 0.2–0.8 mm) was used as the bed material.

The overall net energy efficiency of the integrated system was calculated to assess the system performance of torrefaction, gasification, and power generation:

$$\eta_{net} = \frac{P_{output} - P_{internal}}{P_{input}} \tag{6.3}$$

where P_{output}, $P_{internal}$, and P_{input} are total generated power, internal power consumption, and total energy input (the biomass input (MW)), respectively. The pump and compressor's internal power consumption works in the gasifier and the gas turbine's compressor.

6.1.2 Results and discussion

6.1.2.1 Superheated steam drying and milling performance

The key benefit of using SSD instead of parboiling is that the rice product's consistency does not drop. The SSD process's target moisture content was set to 18 wt% wb at 150°C in this scenario. When the head rice's moisture content was less than 18 wt% wb, particularly at temperatures above 150°C, the yield was significantly reduced. Rice grains are completely gelatinized in the SDD operation, and their physicochemical properties are identical to

Table 6.3 Assumed conditions of the gasification and combined cycle processes.

Properties	Value
Torrefaction	
Temperature (°C)	200–290
Pressure (MPa)	0.1
Flow rate (t/h)	3.6
Husk–straw ratio	4:6
Gasification	
Gasifier type	Dual circulating fluidized bed
Gasifier temperature (°C)	700
Combustor temperature (°C)	850
Average particle size (mm)	0.3–1
Fluidizing medium	Sand (limestone)
Fluidizing medium average diameter (mm)	0.2–0.8
Combustor and gas turbine	
Discharge pressure (MPa)	0.15
Combustor pressure drop (%)	2
Gas turbine isentropic efficiency (%)	90

parboiled rice. The SSD method results based on exergy recovery and the energy consumption for husking and polishing are shown in Table 6.4. The required compressor duty is 0.589 MW, according to the SSD simulation. Heat recovery could be achieved efficiently using exergy elevation and heat coupling. In addition, the outlet steam pressure is 195 kPa, and the compressor temperature is 230°C.

The expected blower work is 5.5 kW because the minimum recycled steam required as a fluidizing agent is 0.2 kg/s. As the flow rate of recycled steam is increased, the blower work increases (Fig. 6.6). Because of the increased contact between the rice grain and the steam, a higher steam flow rate can result in faster and more efficient drying. For further investigation, the impact of raising the recycled steam flow rate, especially on the quality of rice products, needs to be explained.

Fig. 6.7 demonstrates the temperature–enthalpy diagram of the rice grain during SSD. The hot and cold materials are indicated by solid and dashed lines, respectively. The ΔT is the temperature difference between the rice grain and the compressed steam inside the heat exchanger (19°C). Latent heat exchange dominates in this fluidized bed dryer. It occurs

Table 6.4 Results of the superheated steam drying (SSD) performance.

Properties	Value
Compressor performance	
Compressor outlet temperature (°C)	230
Compressor outlet pressure (kPa)	195
Steam condensing temperature (°C)	119
Compression work (MW)	0.589
Dryer	
Mean temperature difference/LMTD (°C)	65
Duty (MW)	8.38
Bed temperature (°C)	150
Drying time (min)	5
Husking and polishing	
Dried grain product (t/d)	118.95
The electrical consumption (MW)	0.22

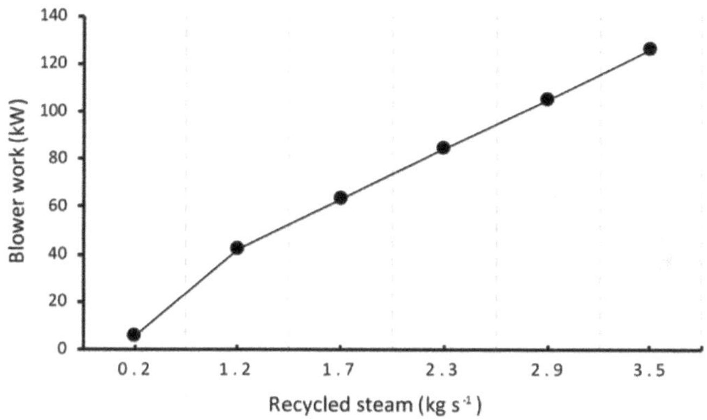

Figure 6.6 Effect of recycled steam on blower work at minimum fluidization velocity U_{mf}.

between the rice grain's water drying heat and the compressed vapor's condensation heat. As a result, the hot stream of compressed vapor curves is nearly parallel to the cold rice grain water streams. It indicates that each form of heat, including sensible and latent heat, has excellent heat coupling. As a result, exergy loss is decreased during the SSD process, and the amount of external energy needed is significantly reduced.

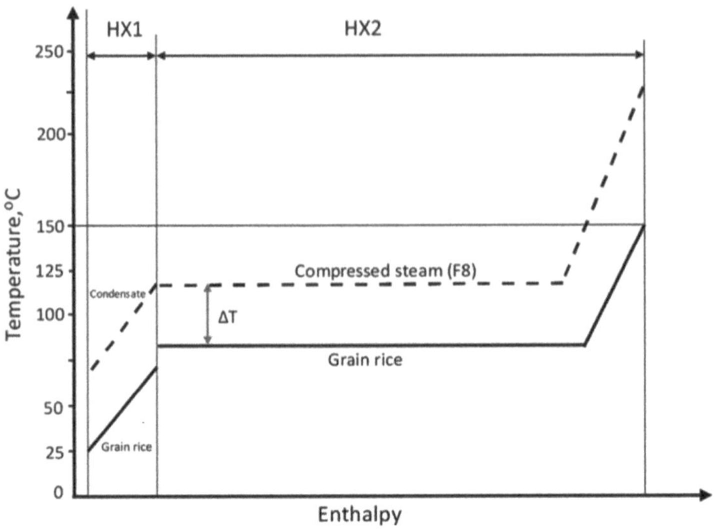

Figure 6.7 Temperature–enthalpy diagram during superheated steam drying (SSD) process of rice grain employing exergy recovery.

6.1.2.2 Comparison with parboiling process

The hydrothermal treatment parboiling is an energy-intensive operation. The parboiling process's energy intensity is primarily affected by the soaking temperature, soaking time, steaming time, and pressure. Rice grain is soaked in water at room temperature until its moisture content exceeds 42–54 wt% wb in the typical parboiling process. This stage takes 36–72 h. After that, the soaked rice grain is steamed at a temperature of 100–105°C. After that, the parboiled rice grain is dried to a moisture content of 16 wt% wb. In comparison to the SSD method proposed in this report, Fig. 6.8 shows the energy consumptions during the parboiling process. The net energy requirements for hot soaking, steaming, and drying for Parboiling 1 are 360, 105.5, and 547 MJ t-parboiled-rice^{-1} (total of 1012.5 MJ/t), according to the International Development Research Center in Canada. Regrettably, the method of parboiling was not outlined in detail. The energy consumption for the parboiling method was recorded as 1680, 1400, and 1659 MJ t-parboiled-rice^{-1}, respectively, by Ahiduzzaman et al. (Parboiling 2) [22], Kirubi et al. (Parboiling 3) [23], and Roy et al. (Parboiling 4) [24]. In Parboiling 2 and 4, the thermal process of direct combustion of rice husk was used to provide the required energy. The SSD with exergy recovery requires a 255.4 MJ t-parboiled-rice^{-1} energy input rate, significantly less than that needed by other parboiling systems. Exergy

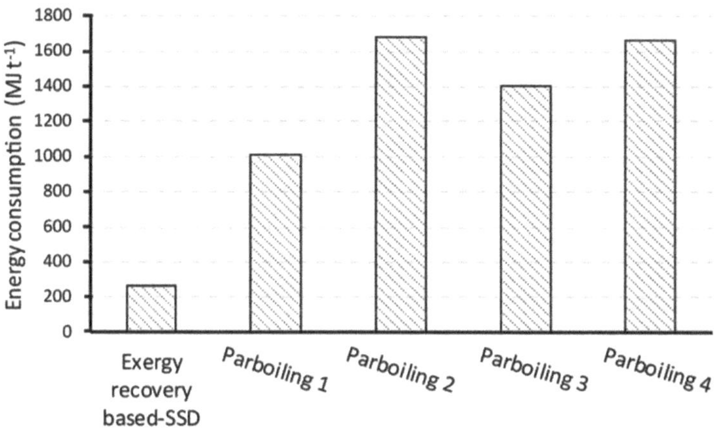

Figure 6.8 Comparison of required energy during the parboiling process and exergy-recovery-based superheated steam drying (SSD).

destruction is the main source of inefficiency in the parboiling process. During the combustion of husks for steam and hot air generation, exergy is lost inexorably. Exergy losses during the parboiling process are often increased by unrecovered steam, heat, or hot water.

6.1.2.3 Performance of torrefaction, gasification, and power generation

The correlation between torrefaction temperature and solid yield of torrefied biomass, energy requirement, and gasification efficiency is depicted in Fig. 6.9. Since more volatiles is removed during the torrefaction process, increasing the torrefaction temperature lowers the solid yield. At 200°C, the torrefaction process is similar to drying but with a higher volume of water removed and a small amount of evaporated volatile matter. A temperature rise in the torrefaction process has little effect on the required energy for torrefaction and gasification efficiency. The needed energy increases marginally from 0.7 to 0.91 MW when the torrefaction temperature is raised from 200 to 290°C. On the other hand, increases in torrefaction temperature can reduce total energy efficiency from 32% to 23% (Fig. 6.10).

Since an increase in torrefaction operation temperature has a significant impact on the solid yield, it can greatly influence the calorific value of the syngas generated in the gasification process. The produced power drops from 4.48 to 3.26 MW as the torrefaction temperature rises from 200 to 290°C. Furthermore, the compressor work in power generation, especially

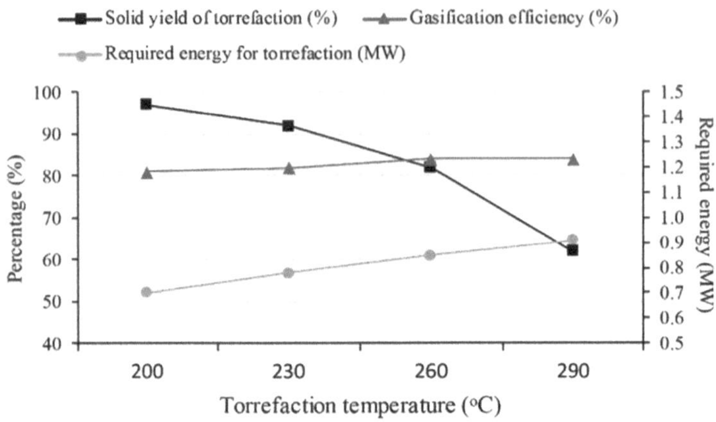

Figure 6.9 Effect of torrefaction temperature on torrefaction and gasification performance.

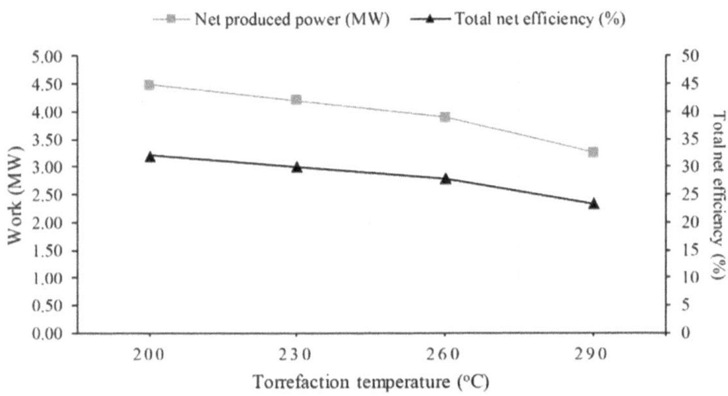

Figure 6.10 Net produced power and total efficiency at different torrefaction temperatures.

in the gas turbine (which consumes more than 90% of the system's internal power), dominates the system's internal power consumption, followed by the blower and water pump. In this analysis, a thermochemical diagram of rice production and power generation is shown in Fig. 6.11. The overall process requires no external heat from the outside to satisfy the internal demand, as seen in the diagram. Since rice is one of the most energy-intensive industries, it is crucial to formulate a strategy for improving energy efficiency. Integration of thermochemical steps can significantly boost the efficiency of future industries, particularly in rice processing.

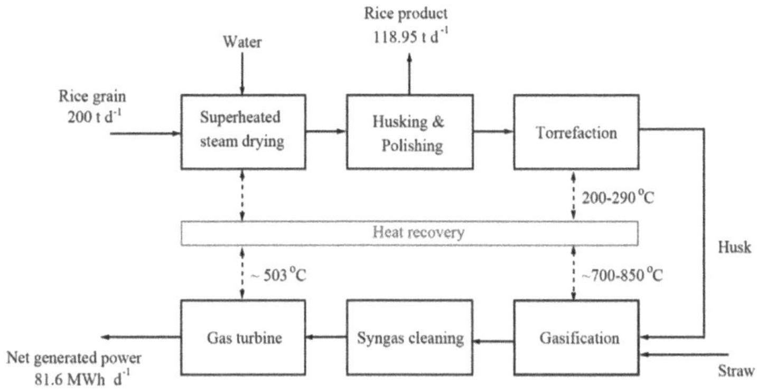

Figure 6.11 Schematic diagram of the thermochemical rice production and electricity generation.

Finally, the proposed method employs exergy recovery to minimize rice production's energy consumption significantly. Rice by-products can also be used to generate electricity on their own to meet the rice production process's needs. According to the calculations, a single rice production system with a factory capacity of 200 t d1 will produce 81.6 MWh d1 of surplus electricity. The synergetic integration of rice production and electricity generation using rice husk and rice straw is critical to improving the long-term energy supply. Since the rice industry uses some of the most energy-intensive manufacturing processes, integrating thermochemical processes could result in significant cost savings and environmental benefits.

6.2 Coal cofiring of hydrothermal-treated empty fruit bunch

Many kinds of research have been carried out in order to find a better way to recover empty fruit bunch (EFB) for energy use, such as cofiring with coal and thermal conversion for solid, liquid, or gaseous biofuel processing. Unfortunately, the high water content of EFB (up to 60% wb) necessitates biomass pretreatment such as drying before burning to result in low overall process performance. The combustion of EFB was investigated by Ninduangdee and Kuprianov using fluidized bed technology and various bed materials [25]. Diego et al. utilized the EFB to perform an experimental analysis on ethanol production [26]. According to their findings, using an alkaline pretreatment for the EFB is not a viable technology. Yan et al.

investigated the effect of coal particle size distribution on the cocombustion efficiency of sewage sludge and coal in terms of cofiring or cocombustion [27]. They found that reducing the fuel particle size improves combustion efficiency. Luk et al. suggested utilizing EFB to combine drying and power generation [28]. Unfortunately, no substantial effort was made to reduce exergy loss in their system, resulting in low energy efficiency. Furthermore, Aziz et al. produced a combined use of EFB and palm oil mill effluent for power generation using a gas engine by gasification and digestion, respectively [29]. While their system appears to be viable for use, it is built as a stand-alone system at a milling site, necessitating the installation of a new gasification system, which raises capital costs.

Given the high potential of EFB and many coal-fired power plants, EFB utilization through HT treatment and cofiring with existing coal-fired power plants and those being built or planned becomes critical. However, studies dealing with the effort to determine the impact of hydrothermally treated EFB (HT-EFB) cofiring to coal-fired combustors are difficult to be found, to the best of the authors' knowledge. In order to estimate the potential for retrofitting an existing power plant to cofiring with HT-EFB, a new approach must be developed quickly. As a result, the research should focus on two key issues to propose efficient EFB use, especially coal cofiring. First, the thermal behaviors of coal cofiring with HT-EFB in a drop tube furnace (DTF) are modeled and analyzed using computational fluid dynamics (CFD), including temperature profiles and exhaust gas composition (CO and CO_2). Our reports have discussed performance for all stages of the combustion, including the combustion temperatures, kinetics behavior, and concentration of the produced gases (CO_2, CO, O_2, NOx, and SOx) [30,31]. The current state-of-the-art CFD modeling-based study can solve the complexity of the interdependent processes such as turbulence, heat transfer via radiation, complex reactions in both the particle and gas phases, and the produced gas. Second, the topic that will be discussed in this chapter, an integrated system model is developed to incorporate CFD simulations with the entire plant processes, such as hydrothermal treatment, combustion, and steam cycles.

6.2.1 Overall system design

The schematic diagram of an integrated coal cofiring process with HT-EFB for electricity generation is illustrated in Fig. 6.12. Material and energy (electricity, heat) flows are expressed by solid and dotted lines, respectively. Before being hydrothermally processed, raw EFB is shredded to a smaller

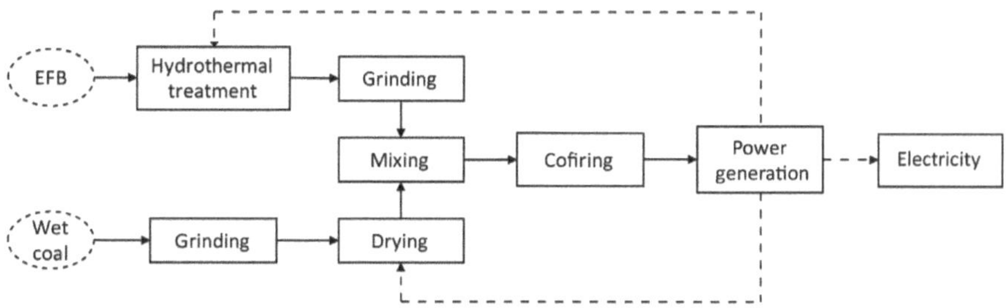

Figure 6.12 Basic schematic diagram of hydrothermally treated EFB (HT-EFB) and coal cofiring system.

size. In general, HT is carried out in the subcritical water region at a temperature of less than 250°C. HT has been studied on a variety of biomasses to generate hydrochar by many researchers. Continuous HT at a slightly higher temperature than saturated is used to remove the drying phase after HT. As a result, the moisture within the EFB evaporates and is expelled along with the steam needed for the next HT process. The HT-EFB that is produced is released from the reactor with low moisture content. Coal, on the other hand, is ground and dried before being combined with HT-EFB to reduce moisture content. The combustor then cofires the mixed HT-EFB and coal, generating high-temperature heat for steam generation through a boiler. The steam is then expanded in the steam engine, which produces electricity. Furthermore, the flue gas from the boiler is primarily used in HT and coal drying.

6.2.1.1 Process integration: process modeling and calculation

In a coal-fired power plant, the main components are drying, combustion, and a power generation system. Before being fed to the dryer unit, the wet coal is ground into smaller particles. In practice, coal particle grinding is done in two stages: before drying and after drying (Table 6.5). The former is necessary to achieve a uniform and small particle size (approximately 1–2 mm) for efficient heat transfer and moisture removal during drying. The latter is designed to reduce the particle size to less than 0.15 mm (pulverized coal) in order to boost combustion performance. The combustor chamber is then fed with dried coal using air as the feeding gas. The heat produced by combustion is recovered by a superheater and economizer and used to produce steam for the steam turbine as it happens within the combustor. No effort is made in this conventional method to reduce exergy destruction or increase overall energy efficiency.

Table 6.5 Material composition.

Component	Raw coal	Dried coal	Raw EFB	HT-EFB
Proximate analysis				
Fixed carbon (wt% wb)	24.93	40.23	3.71	28.62
Volatile matter (wt% wb)	25.76	41.57	34.84	62.57
Moisture (wt% wb)	48.76	17.30	60.00	3.00
Ash (wt% wb)	0.56	0.90	1.46	5.82
Ultimate analysis				
C (wt% wb)	35.30	56.98	17.97	52.92
H (wt% wb)	2.29	3.69	2.49	5.35
O (wt% wb)	11.23	18.13	17.60	32.06
N (wt% wb)	1.75	2.83	0.47	0.85
S (wt% wb)	0.11	0.17	0.01	0.00
Calorific value (MJ/kg)	13.84	22.34	17.02	22.22

Fig. 6.13 depicts a schematic process flow diagram of an integrated system based on enhanced process integration (EPI) concepts that include EFB HT-treatment, coal drying, cofiring, and power generation. The involved processes are integrated using a combination of exergy recovery and process integration. EPI method has been adequately demonstrated and implemented in some areas, including coal conversion, biomass gasification, and algae-based power generation, in order to maximize heat circulation and reduce exergy loss in the integrated system [32]. Before being fed to the HT reactor, raw EFB is shredded to a smaller, more uniform size. The primary reactor for the HT treatment is a continuous autoclave with a tube heat exchanger built inside. For this purpose, the continuous autoclave has already been developed and used in industry. Tube heat exchangers are used to allow heat transfer from the compressed evaporated moisture to the EFB, resulting in self-heat exchange. The HT treatment is carried out at a temperature slightly above saturation; as a result, the moisture in the EFB is in vapor form, and the HT-EFB formed has a low moisture content. As a result, drying after HT treatment can be avoided, and the HT-EFB produced after grinding to produce pulverized HT-EFB can be directly cofired with coal.

Superheated steam with vapor recompression technology is also used to dry coal. Before being fed to the combustor, dried coal and HT-EFB are premixed in a specific mass ratio. Based on the experimental work [33], the sorption isotherm for LRC can be determined, which is needed to measure the relative vapor pressure as a driving force during drying [33].

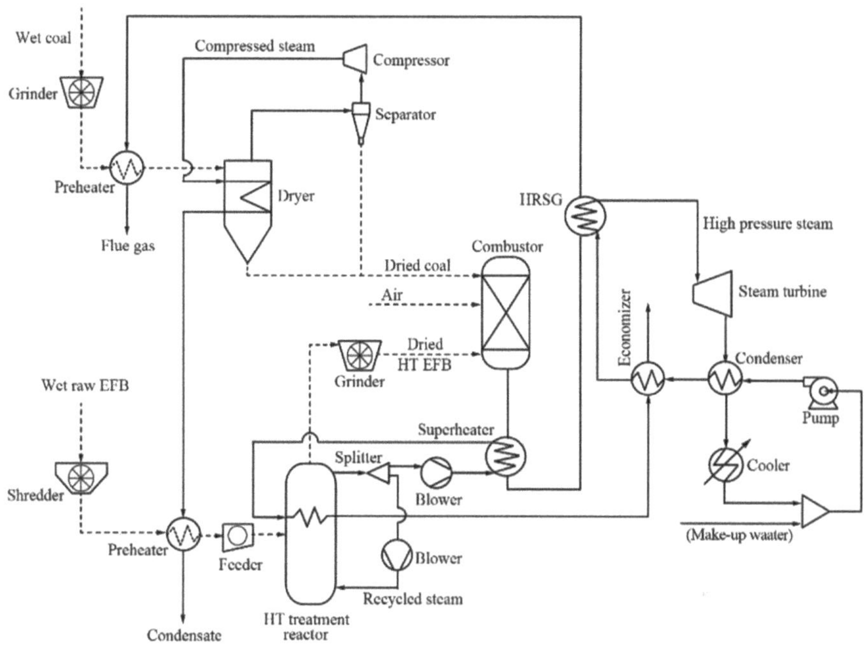

Figure 6.13 Schematic process flow diagram of integrated HT treatment of EFB, coal drying, cofiring, and power generation.

$$\frac{p}{p_{sat}} = 1 - exp\left[- 2.53\left(T_{dry} - 273\right)^{0.47}\left(\frac{MC}{100 - MC}\right)^{1.58}\right] \tag{6.4}$$

Table 6.6 indicates the other assumed conditions. The calculation is carried out with the aid of the process simulator Aspen Plus version 8.8. (Aspen Technology, Inc.). Raw EFB and wet coal flow concentrations are 12.5 and 35 kg/s, respectively. As a result, the mass fraction of HT-EFB during cofiring is about 19.5%.

The following equation obtains the net power generation efficiency (η),

$$\eta = \frac{P_{gross} - P_{int}}{\left(\dot{m}_{coal} \times CV_{coal}\right) + \left(\dot{m}_{EFB} \times CV_{EFB}\right)} \tag{6.5}$$

where P_{gross}, P_{int}, $\dot{m}_{coal}$, $\dot{m}_{EFB}$, CV_{coal}, and CV_{EFB} are the power output (gross generated power), internal energy consumption (compressor work, pump, blower, etc.), the mass flow rate of coal, the mass flow rate of EFB, the calorific value of wet coal, and calorific value of the raw EFB, respectively.

Table 6.6 Assumptions given during the calculation of the integrated system.

Parameter(s)	Value
Adiabatic efficiency of compressor and blower (%)	87
The minimum approach of heat exchangers (°C)	10
HRSG outlet pressure (MPa)	20
Steam turbine isentropic efficiency (%)	90
Minimum outlet vapor quality (%)	0.9
The discharge pressure of the steam turbine (MPa)	0.13

6.2.2 Result and discussion

Fig. 6.14 contrasts energy consumed during coal drying, HT treatment of EFB, and internal consumption in the power generation system. Table 6.7 summarizes the findings on a power plant's performance. The power generation's internal consumption, such as the high-pressure pump, reflects the highest power consumption during the cycle. When it comes to these auxiliaries, the water pumping for the steam cycle and the air blower for combustion use the most energy (about 70% of total). The proposed coal drying and HT treatment of the EFB, on the other hand, result in lower energy consumption due to very low exergy loss.

The proposed integrated system with EPI, which includes the processes for coal drying and HT treatment of the EFB, will achieve a net power generation efficiency of 38.9%. Fig. 6.15 depicts the simulation's major findings. The integrated system's high net power generation performance can be due to substantial energy savings in coal drying and HT treatment of

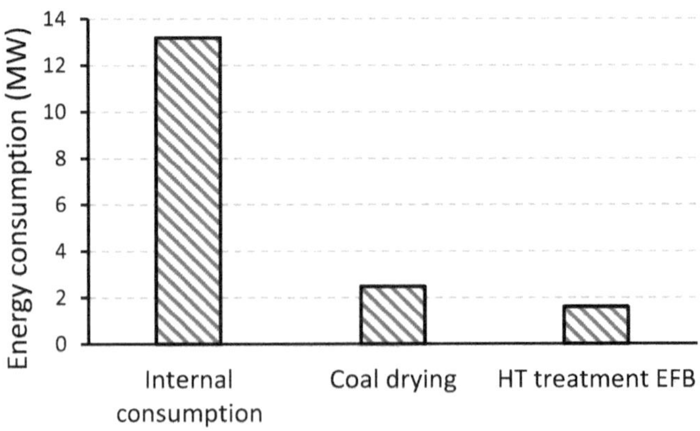

Figure 6.14 Comparison of the energy demand in coal drying, HT treatment of EFB, and internal consumption for power generation.

Table 6.7 Calculation result.

Steam cycle data	Value
The mass flow rate of fuel	26.9 kg/s
Biomass to fuel ratio	19.5%
Fuel input energy	334 MW
Steam temperature	525°C
Steam pressure	20 MPa
Steam flow rate	702-Ton/hr
Outlet vapor quality	92%
Flue gas temperature (after HRSG)	120°C
Gross power output	147.9 MW
Net power output	129.9
Auxiliary power ratio (including drying and hydrothermal process)	12%
Net efficiency	~38.9%

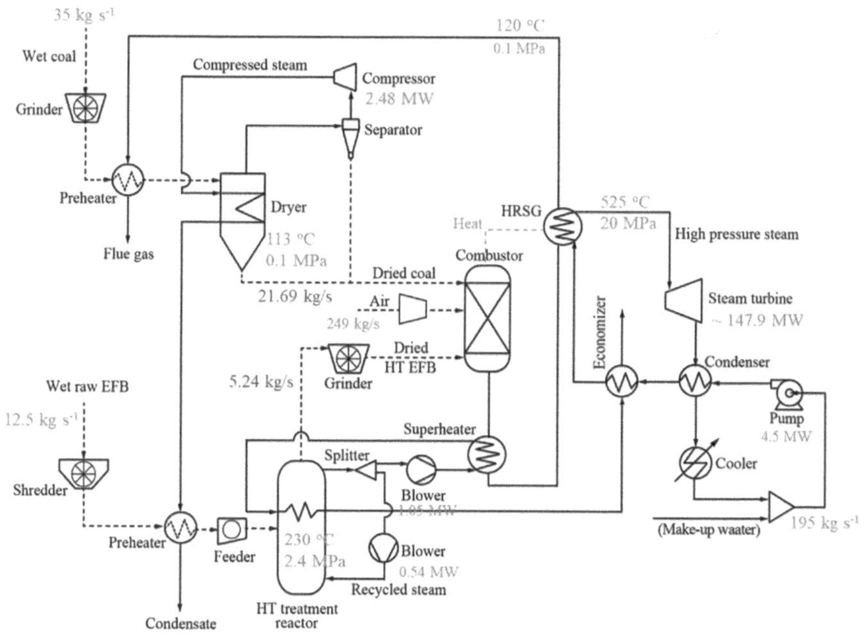

Figure 6.15 Overall results of the simulation (some key parameters).

the EFB. It is clear from Fig. 6.13 that changes are needed primarily in the drying and hydrothermal treatment processes. The conceptual model can be used and replicated in a variety of capacities. EPI technology, which incorporates exergy recovery and process integration, is used in the suggested scheme. EPI, on the other hand, uses a methodological approach that aims to integrate both technologies effectively. As a result, they can be used easily and effectively to achieve the highest possible energy efficiency. However, retrofitting the drying system, hydrothermal treatment system, and boiler would entail additional investment costs.

More attention should be paid to the materials, thermal stresses during startup, and load fluctuations in high-temperature heat exchangers, especially for the hydrothermal treatment process. Materials such as carbon fiber-SiC composites or different ceramic oxides may be used to account for the heat exchanger's high temperatures. However, the use of these materials can result in a higher cost for a heat exchanger that can withstand temperatures above 700°C. Furthermore, improving both the drying and hydrothermal treatment systems' design and operation will result in higher efficiencies.

6.3 Conclusion

The use of EFB as an alternative energy source has been suggested and investigated. Initially, the activity of coal cofiring with HT-EFB should be modeled and evaluated using CFD analysis at various mixing mass fractions. In this report, an advanced integrated system that included HT treatment for the EFB, coal drying, cofiring, and power generation was modeled. It is possible to achieve very low energy consumption during coal drying and HT treatment of the EFB, which are 2.48 and 1.6 MW, respectively. Finally, when coal drying and HT treatment of the EFB are added, the proposed integrated system's net power generation efficiency is about 38.9%. The proposed system's implementation is expected to increase EFB's efficient usage, minimize environmental impact, and prolong existing coal-fired power plants' lives.

References

[1] Prakash Chandak S, Ranga Chari K, Ahmed Memon M. Converting waste agricultural biomass into energy: experiences and lessons learnt from a capacity building and technology demonstration project in India. [n.d].

[2] Rosales-Calderon O, Arantes V. A review on commercial-scale high-value products that can be produced alongside cellulosic ethanol. Biotechnol Biofuels 2019;12:240. https://doi.org/10.1186/s13068-019-1529-1.

[3] Alhazmi H, Loy ACM. A review on environmental assessment of conversion of agriculture waste to bio-energy via different thermochemical routes: current and Future Trends. Bioresour Technol Rep 2021:100682. https://doi.org/10.1016/j.biteb.2021.100682.

[4] Yusof M, Farid N, Zainal Z, Azman M. Characterization of rice husk for cyclone gasifier. Appl Sci 2008:622–8.
[5] United Nations Environmental Programme Division of Technology. Industry and economics international environmental technology centre Osaka/Shiga J. Converting waste agricultural biomass into a resource. United Nations Environ Program 2009:1–437.
[6] Shafie SM, Shafie SM, Shafie SM. A review on paddy residue based power generation: energy, environment and economic perspective. Renew Sustain Energy Rev 2016;59:1089–100. https://doi.org/10.1016/j.rser.2016.01.038.
[7] Kwofie EM, Ngadi M. A review of rice parboiling systems, energy supply, and consumption. Renew Sustain Energy Rev 2017;72:465–72. https://doi.org/10.1016/j.rser.2017.01.014.
[8] Taechapairoj C, Dhuchakallaya I, Soponronnarit S, Wetchacama S, Prachayawarakorn S. Superheated steam fluidised bed paddy drying. J Food Eng 2003;58:67–73. https://doi.org/10.1016/S0260-8774(02)00335-7.
[9] Wu J, McClements DJ, Chen J, Hu X, Liu C. Improvement in nutritional attributes of rice using superheated steam processing. J Funct Food 2016;24:338–50. https://doi.org/10.1016/j.jff.2016.04.019.
[10] Suramaythangkoor T, Gheewala SH. Potential of practical implementation of rice straw-based power generation in Thailand. Energy Pol 2008;36:3183–7. https://doi.org/10.1016/j.enpol.2008.05.002.
[11] Yoon SJ, Son YI, Kim YK, Lee JG. Gasification and power generation characteristics of rice husk and rice husk pellet using a downdraft fixed-bed gasifier. Renew Energy 2012;42:163–7. https://doi.org/10.1016/j.renene.2011.08.028.
[12] Calvo LF, Gil MV, Otero M, Morán A, García AI. Gasification of rice straw in a fluidized-bed gasifier for syngas application in close-coupled boiler-gasifier systems. Bioresour Technol 2012;109:206–14. https://doi.org/10.1016/j.biortech.2012.01.027.
[13] Prabowo B, Umeki K, Yan M, Nakamura MR, Castaldi MJ, Yoshikawa K. CO2-steam mixture for direct and indirect gasification of rice straw in a downdraft gasifier: laboratory-scale experiments and performance prediction. Appl Energy 2014;113:670–9. https://doi.org/10.1016/j.apenergy.2013.08.022.
[14] Chiang KY, Lin YX, Lu CH, Chien KL, Lin MH, Wu CC, et al. Gasification of rice straw in an updraft gasifier using water purification sludge containing Fe/Mn as a catalyst. Int J Hydrogen Energy 2013;38:12318–24. https://doi.org/10.1016/j.ijhydene.2013.07.041.
[15] Jenkins BM, Baxter LL, Miles TR, Miles TR. Combustion properties of biomass. Fuel Process Technol 1998;54:17–46. https://doi.org/10.1016/S0378-3820(97)00059-3.
[16] Demirbas A. Combustion characteristics of different biomass fuels. Prog Energy Combust Sci 2004;30:219–30. https://doi.org/10.1016/j.pecs.2003.10.004.
[17] Aziz M, Prawisudha P, Prabowo B, Budiman BA. Integration of energy-efficient empty fruit bunch drying with gasification/combined cycle systems. Appl Energy 2015;139:188–95. https://doi.org/10.1016/j.apenergy.2014.11.038.
[18] Taechapairoj C, Prachayawarakorn S, Soponronnarit S. Characteristics of rice dried in superheated-steam fluidized-bed. Dry Technol 2004;22:719–43. https://doi.org/10.1081/DRT-120034259.
[19] Goyal SK, Jogdand SV, Agrawal AK. Energy use pattern in rice milling industries—a critical appraisal. J Food Sci Technol 2012;51:2907–16. https://doi.org/10.1007/s13197-012-0747-3.
[20] Goyal S, Jogand S, Agarwal A. A study of energy audit in rice processing machines. Progress Agric 2008;8:34–8.
[21] Chen D, Zhou J, Zhang Q, Zhu X, Lu Q. Upgrading of rice husk by torrefaction and its influence on the fuel properties. BioResources 2014;9:5893–905. https://doi.org/10.15376/biores.9.4.5893-5905.

[22] Ahiduzzaman M, Sadrul Islam AKM. Energy utilization and environmental aspects of rice processing industries in Bangladesh. Energies 2009;2:134–49. https://doi.org/10.3390/en20100134.
[23] Kirubi C, Jacobson A, Kammen DM, Mills A. Community-based electric micro-grids can contribute to rural development: evidence from Kenya. World Dev 2009;37:1208–21. https://doi.org/10.1016/j.worlddev.2008.11.005.
[24] Roy P, Shimizu N, Shiina T, Kimura T. Energy consumption and cost analysis of local parboiling processes. J Food Eng 2006;76:646–55. https://doi.org/10.1016/j.jfoodeng.2005.06.034.
[25] Ninduangdee P, Kuprianov VI. A study on combustion of oil palm empty fruit bunch in a fluidized bed using alternative bed materials: performance, emissions, and time-domain changes in the bed condition. Appl Energy 2016;176:34–48. https://doi.org/10.1016/j.apenergy.2016.05.063.
[26] Piarpuzán D, Quintero JA, Cardona CA. Empty fruit bunches from oil palm as a potential raw material for fuel ethanol production. Biomass Bioenergy 2011;35:1130–7. https://doi.org/10.1016/j.biombioe.2010.11.038.
[27] Yan YF, Zhang ZE, Zhang L, Zhang L. Influence of coal properties on the co-combustion characteristics of low-grade coal and city mud. Glob Nest J 2014;16:330–9. https://doi.org/10.30955/gnj.001315.
[28] Luk HT, Lam TYG, Oyedun AO, Gebreegziabher T, Hui CW. Drying of biomass for power generation: a case study on power generation from empty fruit bunch. Energy 2013;63:205–15. https://doi.org/10.1016/j.energy.2013.10.056.
[29] Aziz M, Kurniawan T, Oda T, Kashiwagi T. Advanced power generation using biomass wastes from palm oil mills. Appl Therm Eng 2017;114:1378–86. https://doi.org/10.1016/j.applthermaleng.2016.11.031.
[30] Darmawan A, Budianto D, Tokimatsu K, Aziz M. Analysis of biomass waste cofiring into existing coal-fired power plant using computational fluid dynamics. Comput. Fluid Dyn. - Basic Instruments Appl. Sci., Intech 2018:217–36. https://doi.org/10.5772/intechopen.70561.
[31] Darmawan A, Budianto D, Aziz M, Tokimatsu K. Retrofitting existing coal power plants through cofiring with hydrothermally treated empty fruit bunch and a novel integrated system. Appl Energy 2017;204:1138–47. https://doi.org/10.1016/j.apenergy.2017.03.122.
[32] Aziz M. Power generation from algae employing enhanced process integration technology. Chem Eng Res Des 2016;109:297–306. https://doi.org/10.1016/j.cherd.2016.02.002.
[33] Allardice DJ, Evans DG. The-brown coal/water system: Part 2. Water sorption isotherms on bed-moist Yallourn brown coal. Fuel 1971;50:236–53. https://doi.org/10.1016/0016-2361(71)90014-7.

CHAPTER 7

Exergoeconomic, exergoenvironmental, and conclusion

Arif Darmawan[1] and Muhammad Aziz[2]
[1]Agency for the Assessment and Application of Technology (BPPT), Puspiptek Serpong, Tangerang Selatan, Indonesia; [2]Institute of Industrial Science, The University of Tokyo, Meguro-ku, Tokyo, Japan

The high performance of the proposed integrated systems should be feasible technically, economically, and environmentally. In the economic aspect, the investment needed for a complete unit would be largely influenced by investment in modifications, including the redesign of a few reactors. Given that our models include new technology without operating full-scale plants, the investment cost remains difficult to ascertain. For example, for an integrated gasification combined cycle (IGCC) system, the amount of investment needed is estimated to be about 60%–90% higher than that required for a conventional combustion system using the steam power cycle. Inevitably, to elaborate and understand the economic viability, more research is necessary. The thermodynamics-based method and findings used in this book may prove useful for further economic evaluation, particularly for developing a real process or commercial application. Exergoeconomic analysis can be used in thermal systems as a modern tool for integrating exergy and economic analysis to investigate biomass-based energy conversion processes. Financial caused by irreversibilities (exergy losses) are examined, including correlations between these costs and the investment and operating costs for each part of the plant. Furthermore, as a complement, the exergoenvironmental analysis can be adopted to assign environmental impacts within each process component for both energy and material flows as well as thermodynamic inefficiencies [1,2].

7.1 Exergoeconomic and exergoenvironmental analysis

Exergoeconomics is an engineering branch that combines thermodynamic-based exergy analysis and economic principles to design a cost-effective system for sustainable development. The general term of thermoeconomics

Innovative Energy Conversion from Biomass Waste
ISBN 978-0-323-85477-1
https://doi.org/10.1016/B978-0-323-85477-1.00002-6

is more fitting according to some suggestions if exergy costing is not applied. For each flow in a system, a parameter called flow cost rate is defined, and a cost balance is written for each component k as,

$$\dot{C}_{q,k} + \sum_i \dot{C}_{i,k} + \dot{Z}_k = \sum_e \dot{C}_{e,k} + \dot{C}_{w,k} \tag{7.1}$$

In the equation, $\dot{C}$ denotes the price of exergy flow or flow cost rate and $\dot{Z}$ denotes capital cost rate. The value (Z) is a function of the annual operational period, system life, interest, and escalation. The q, i, e, and w are the heat, interest rate, exit condition, and work, respectively. For a system operating at a steady state, there may be several entering and exiting mass streams and interactions of heat and work with the surroundings. There are common assumptions adopted during the exergoeconomics analysis as follows:

- The system operates at a steady-state condition.
- Kinetic and potential energy effects are ignored.
- Ideal gas principles are used.
- The exergy output produced by the reactor is at chemical equilibrium.
- Heat losses such as due to separation and mass distribution from the components are neglected.
- The reference environment temperature and pressure are 20–25°C and 101.325 kPa, respectively.

In this analysis, the fuel/feedstock and product exergy need to be carefully identified. The product exergy is specified by the component under consideration, whereas the fuel is the consumed feedstock. Also, the optimization of thermodynamic-based exergy will differ from exergoeconomic optimization. Exergoeconomic optimization focuses on exergy and economic performance. It aims to reduce costs, including operating costs associated with all process flows, capital expenditures, and internal and external exergy losses [3]. As a result, the overall cost of exergoeconomic optimization is typically lower than the corresponding cost of thermodynamic-based optimization.

The exergoenvironmental analysis consists of three key steps: conversion process exergy analysis, life cycle assessment (LCA), and assigning of environmental impacts to the exergy streams in the process. Exergoenvironmental variables are subsequently calculated, and the exergoenvironmental assessment is carried out. Possibilities for enhancing environmental performance can be established based on analyzing the

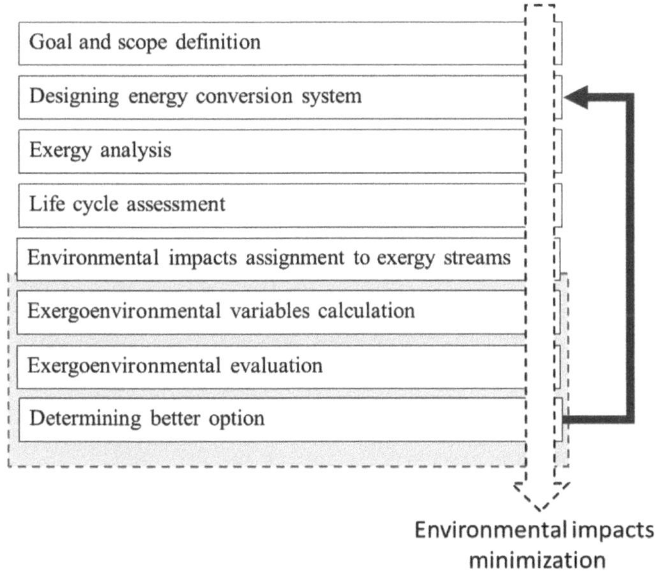

Figure 7.1 Structure of the exergoenvironmental method. *(Adapted from Meyer L, Tsatsaronis G, Buchgeister J, Schebek L. Exergoenvironmental analysis for evaluation of the environmental impact of energy conversion systems. Energy 2009;34:75–89. https://doi.org/10.1016/j.energy.2008.07.018.)*

process and its components (see Fig. 7.1). Here, the assignment of environmental analysis results to exergy streams is performed in analogy to the assignment of costs to exergy streams in exergoeconomics [4].

Depending on the system or component being analyzed, it may be useful to distinguish between physical (PH) and chemical (CH) exergy. In this case, a specific environmental impact for every exergy component must be known to calculate the environmental impact rate $\dot{B}_j$ or the average specific environmental impact b_j:

$$\dot{B}_j = \dot{B}_j^{PH} + \dot{B}_j^{CH} = b_j^{PH}\dot{E}_j^{PH} + b_j^{CH}\dot{E}_j^{CH}$$

The environmental impact rates associated with heat $\dot{Q}$ and work $\dot{W}$ transfers are calculated as follows:

$$\dot{B}_q = b_q\dot{E}_q$$
$$\dot{B}_w = b_w\dot{W}$$

The exergy rate associated with a heat transfer is calculated using the following equation:

$$\dot{E}_q = \left(1 - \frac{T_0}{T_j}\right)\dot{Q}$$

where T_0 and T_j denote the ambient temperature and the temperature at which the heat transfer crosses the system's boundary, respectively. It is usually assumed that all heat transfers to the environment take place at $T_j = T_0$. Therefore, the environmental impact balance for the kth component states that the sum of environmental impacts associated with all input streams plus the component-related environmental impact is equal to the sum of the environmental impacts associated with all output streams,

$$\sum_{j=1}^{n} \dot{B}_{j,k,\text{in}} + \dot{Y}_k = \sum_{j=1}^{n} \dot{B}_{j,k,\text{out}}$$

Though some researchers employ exergoeconomic analysis and exergoenvironmental analysis in design energy systems, they mostly focus on only one of the two aspects (either exergoeconomic analysis or exergoenvironmental analysis). However, the systems operating under optimized exergoeconomic conditions might not agree with the best exergoenvironmental requirements. "Many thermodynamic systems have bad behavior in the balance of system economic performance and system environmental impacts. The system often has a great economic output while has serious environmental impacts that might need much more capital input to fix. Thus, it is of great significance to carry out exergoeconomic analysis and exergoenvironmental analysis on a thermodynamic system [6]".

7.2 Summary of the book, limitations, and the main conclusion

This book focused on integrated process modeling and improving biomass waste utilization efficiency, as illustrated in Fig. 7.2. Once we address biomass utilization to reduce fossil fuel use, choosing the suitable conversion technology is the key. Different conversion technologies are available, such as thermochemical conversion, biological, chemical, and biochemical. Thermochemical conversion is preferred among these technologies due to more excellent technological performance including combustion, gasification, and

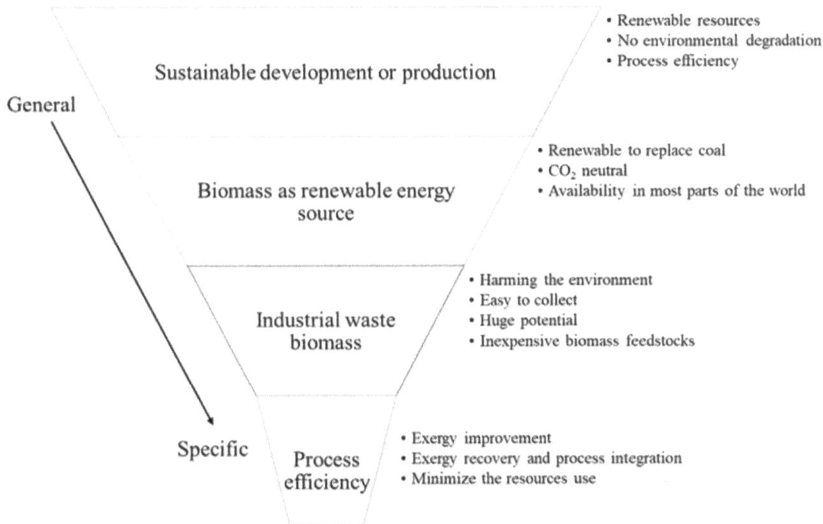

Figure 7.2 Sustainable energy development.

pyrolysis. Nevertheless, several technologies remain under development, such as biomass gasification. Evaluating these conversion technologies' efficiency needs appropriate methods and metrics to qualitatively and quantitatively measure technology's progress. Efficiency evaluation can highlight system imperfections to improve their performance. The efficiency also addresses some of the following questions:

- Does the production of a given product by method A utilize fewer biomass resources than method B?
- What is the biomass resource required (including materials and energy) to make a given product?

High-potential biomass residues from industry such as empty fruit bunch (EFB) or black liquor must be utilized with the appropriate technology to optimize its economic benefit and minimize the environmental impacts. Utilizing biomass waste is advantageous due to feedstock's availability inside the processing mill, thus reducing transportation costs. Different conversion routes from the selected biomass residues including black liquor, empty fruit bunch, and rice waste are proposed and simulated. The integrated systems are proposed by enhanced process integration by employing exergy recovery and process integration technologies to minimize exergy loss. If managed efficiently, industrial waste biomass can replace the conventional utilization of currently adopted carbon-based fuels.

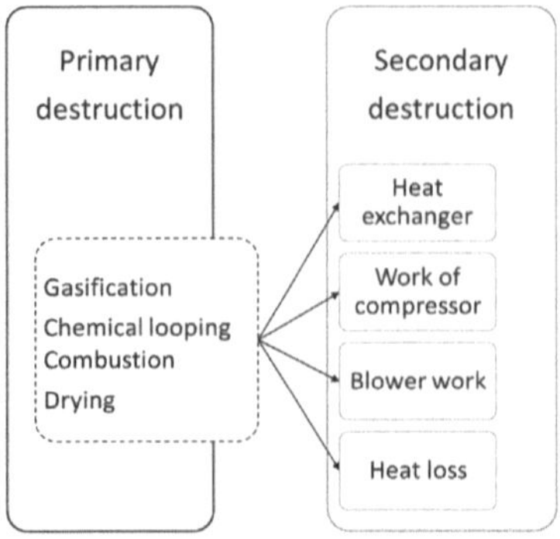

Figure 7.3 Exergy destruction.

Minimization of exergy loss while designing the process is key in improving total energy efficiency. Our discussions show that exergy loss or destruction can be divided into primary and secondary exergy destruction, as shown in Fig. 7.3. Primary exergy destruction is the inevitable exergy loss due to the process's progress or reaction. Secondary exergy destruction is exergy loss due to the surrounding process supporting the progress of the main process; due to heat exchange, work, or heat loss. As heat loss is difficult to define, this study focuses mainly on the exergy destruction due to heat exchange. The irreversibility due to heat exchange in the overall process increases exergy destruction, and the energy efficiency decreases accordingly.

Simulations conducted in this book only propose possibilities and never certificate whether the proposal is feasible or not for the real application/implementation. Many aspects should be considered, such as economic or cost analysis, material availability, plant location, etc., to assess the feasibility (Fig. 7.4). The idea of modeling is to create a model that displays the behavior/operation of the actual process or system. Even if we get some experimental data to support our model, it is still a gap to implement to the real application, including scaling up, design, or material issues. Usually, the experimental study needs to be clarified with a pilot scale. Some aspects are not included in this research, like heat loss as an unknown variable. It is difficult to define the heat loss since many things need to be addressed, e.g., reactor volume, design, size, insulator quality. Some notes during the simulation and modeling include the following:

- Mass/exergy loss during separation and distribution is neglected.

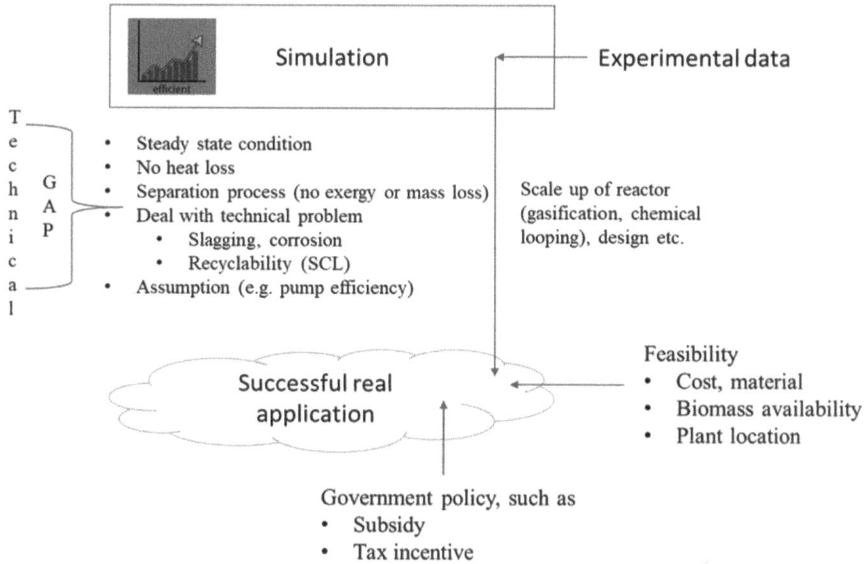

Figure 7.4 Overall comparison between simulation environment and real application.

- Deal with a technical problem such as slagging, corrosion, material recyclability (during SCL)
- Assumptions (pump efficiency, compressor efficiency).

7.3 Main conclusion

- Exergy analysis is a useful tool to evaluate a system's performance.
- There are two engineering design problems in integrated processes. The first is the problem of unit operation design and the second is designing total systems. This book mostly addresses the second issue.
- Ingenuity and creativity are required to design a process and implement efficiency improvement.
- However, comprehensive evaluations are needed considering the interrelations among exergy efficiency itself, economy, and environment for a sustainable system

References

[1] Tsatsaronis G, Winhold M. Exergoeconomic analysis and evaluation of energy-conversion plants-I. A new general methodology. Energy 1985;10:69–80. https://doi.org/10.1016/0360-5442(85)90020-9.

[2] Dincer I, Rosen MA. Exergy: energy, environment and sustainable development. Elsevier Science; 2012.
[3] Ptasinski KJ. Efficiency of biomass energy - an exergy approach to biofuels. Power Biorefin 2017;89. https://doi.org/10.1002/cite.201770067.
[4] Bejan A, Tsatsaronis G, Moran MJ. Thermal design and optimization. Wiley; 1995.
[5] Meyer L, Tsatsaronis G, Buchgeister J, Schebek L. Exergoenvironmental analysis for evaluation of the environmental impact of energy conversion systems. Energy 2009;34:75–89. https://doi.org/10.1016/j.energy.2008.07.018.
[6] Wang W, Wang J, Lu Z, Wang S. Exergoeconomic and exergoenvironmental analysis of a combined heating and power system driven by geothermal source. Energy Convers Manag 2020;211:112765. https://doi.org/10.1016/j.enconman.2020.112765.

Index

Note: 'Page numbers followed by "f" indicate figures and "t" indicate tables.'

C

FSC
www.fsc.org
MIX
Papier | Fördert gute Waldnutzung
FSC® C083411

Kolibri 360 GmbH
Ferdinand-Jühlke-Straße 7
99095 Erfurt, Deutschland
produktsicherheit@kolibri360.de